Functional Photocatalysts: Material Design, Synthesis and Applications

Functional Photocatalysts: Material Design, Synthesis and Applications

Editor

Lin Ju

Basel • Beijing • Wuhan • Barcelona • Belgrade • Novi Sad • Cluj • Manchester

Editor
Lin Ju
School of Physics and
Electric Engineering
Anyang Normal University
Anyang
China

Editorial Office
MDPI
St. Alban-Anlage 66
4052 Basel, Switzerland

This is a reprint of articles from the Special Issue published online in the open access journal *Molecules* (ISSN 1420-3049) (available at: www.mdpi.com/journal/molecules/special_issues/G1ORMYGKA1).

For citation purposes, cite each article independently as indicated on the article page online and as indicated below:

Lastname, A.A.; Lastname, B.B. Article Title. *Journal Name* **Year**, *Volume Number*, Page Range.

ISBN 978-3-7258-1252-3 (Hbk)
ISBN 978-3-7258-1251-6 (PDF)
doi.org/10.3390/books978-3-7258-1251-6

© 2024 by the authors. Articles in this book are Open Access and distributed under the Creative Commons Attribution (CC BY) license. The book as a whole is distributed by MDPI under the terms and conditions of the Creative Commons Attribution-NonCommercial-NoDerivs (CC BY-NC-ND) license.

Contents

About the Editor vii

Lin Ju
Functional Photocatalysts: Material Design, Synthesis and Applications
Reprinted from: *Molecules* **2024**, *29*, 1146, doi:10.3390/molecules29051146 1

Cheng Zuo, Qian Su and Lei Yu
Research Progress in Composite Materials for Photocatalytic Nitrogen Fixation
Reprinted from: *Molecules* **2023**, *28*, 7277, doi:10.3390/molecules28217277 8

Zhimin Yuan, Xianglin Zhu and Zaiyong Jiang
Recent Advances of Constructing Metal/Semiconductor Catalysts Designing for Photocatalytic CO_2 Hydrogenation
Reprinted from: *Molecules* **2023**, *28*, 5693, doi:10.3390/molecules28155693 25

Qian Su, Cheng Zuo, Meifang Liu and Xishi Tai
A Review on Cu_2O-Based Composites in Photocatalysis: Synthesis, Modification, and Applications
Reprinted from: *Molecules* **2023**, *28*, 5576, doi:10.3390/molecules28145576 40

Li-Hua Wang and Xi-Shi Tai
Synthesis, Structural Characterization, Hirschfeld Surface Analysis and Photocatalytic CO_2 Reduction Activity of a New Dinuclear Gd(III) Complex with 6-Phenylpyridine-2-Carboxylic Acid and 1,10-Phenanthroline Ligands
Reprinted from: *Molecules* **2023**, *28*, 7595, doi:10.3390/molecules28227595 65

Madina Bissenova, Arman Umirzakov, Konstantin Mit, Almaz Mereke, Yerlan Yerubayev and Aigerim Serik et al.
Synthesis and Study of $SrTiO_3/TiO_2$ Hybrid Perovskite Nanotubes by Electrochemical Anodization
Reprinted from: *Molecules* **2024**, *29*, 1101, doi:10.3390/molecules29051101 74

Zhi Wang, Changmin Shi, Pengfei Li, Wenzhu Wang, Wenzhen Xiao and Ting Sun et al.
Optical and Photocatalytic Properties of Cobalt-Doped $LuFeO_3$ Powders Prepared by Oxalic Acid Assistance
Reprinted from: *Molecules* **2023**, *28*, 5730, doi:10.3390/molecules28155730 82

Lin Ju, Xiao Tang, Jingli Li, Hao Dong, Shenbo Yang and Yajie Gao et al.
Armchair Janus WSSe Nanotube Designed with Selenium Vacancy as a Promising Photocatalyst for CO_2 Reduction
Reprinted from: *Molecules* **2023**, *28*, 4602, doi:10.3390/molecules28124602 96

Chao Zhang, Jiangwei Xu, Huaizhi Song, Kai Ren, Zhi Gen Yu and Yong-Wei Zhang
Achieving Boron–Carbon–Nitrogen Heterostructures by Collision Fusion of Carbon Nanotubes and Boron Nitride Nanotubes
Reprinted from: *Molecules* **2023**, *28*, 4334, doi:10.3390/molecules28114334 108

Zhaoming Huang, Kai Ren, Ruxin Zheng, Liangmo Wang and Li Wang
Ultrahigh Carrier Mobility in Two-Dimensional IV–VI Semiconductors for Photocatalytic Water Splitting
Reprinted from: *Molecules* **2023**, *28*, 4126, doi:10.3390/molecules28104126 119

Lin Ju, Xiao Tang, Yixin Zhang, Xiaoxi Li, Xiangzhen Cui and Gui Yang
Single Selenium Atomic Vacancy Enabled Efficient Visible-Light-Response Photocatalytic NO Reduction to NH_3 on Janus WSSe Monolayer
Reprinted from: *Molecules* **2023**, *28*, 2959, doi:10.3390/molecules28072959 . . . . **129**

About the Editor

Lin Ju

Ju Lin graduated from the School of Physics of Shandong University, majoring in Microelectronics, with a Bachelor's degree of Science in June 2009, and the School of Physics of Shandong University, majoring in Condensed Matter Physics, with a Doctor's degree of Science in June 2014. In January 2021, he was appointed as a Special Term Professor of the school of Anyang Normal University. As the first author, he has published 24 SCI papers in *Nat. Commun., J. Am. Chem. Soc.* and *J. Mater. Chem. A*, etc. His papers have been cited over 2000 times, and his H-index is 21.

Editorial

Functional Photocatalysts: Material Design, Synthesis and Applications

Lin Ju

School of Physics and Electric Engineering, Anyang Normal University, Anyang 455000, China; julin@aynu.edu.cn

Citation: Ju, L. Functional Photocatalysts: Material Design, Synthesis and Applications. *Molecules* **2024**, *29*, 1146. https://doi.org/10.3390/molecules29051146

Received: 31 January 2024
Accepted: 1 March 2024
Published: 5 March 2024

Copyright: © 2024 by the author. Licensee MDPI, Basel, Switzerland. This article is an open access article distributed under the terms and conditions of the Creative Commons Attribution (CC BY) license (https://creativecommons.org/licenses/by/4.0/).

Rapid industrial and economic growth, experienced on a global scale, has been greatly facilitated by the extensive use and exploitation of traditional energy resources. However, this progress has also resulted in significant environmental pollution and energy scarcity issues, which pose serious threats to our living environment and the future advancement of humanity. Consequently, many researchers are now shifting their focus towards exploring and developing new clean energy solutions, with the aim of addressing the escalating concerns of environmental degradation and energy deficits [1–4]. Among these solutions, solar energy stands out as an endless and renewable power source. Its further application in tackling the current environmental crises could represent a remarkable step forward for the collective well-being of humanity worldwide. Therefore, the progress of solar energy technology carries profound implications [5–7]. Semiconductor photocatalysis is a defined process that utilizes solar energy to initiate excitation in semiconductors, generating photogenerated carriers. These carriers subsequently promote catalytic redox reactions on the surface of the semiconductor. This innovative approach not only enables the transformation and storage of solar energy but also opens possibilities for conversion into various forms of energy [8–11].

In this Special Issue, our focus is centered on the design, synthesis, and diverse applications of functional photocatalysts. We have curated a collection of ten high-quality papers that are poised to capture the interest of researchers in the field of photocatalysis. Among them, four papers present a comprehensive examination of CO_2 conversion methods, two explore the synthesis of NH_3, one is dedicated to investigating the water splitting process, another delves into the degradation of organic pollutants, while the final two papers focus on the development of promising heterojunction photocatalysts. We hope that this Special Issue will act as a catalyst for further advancements in photocatalytic technologies, spurring progress in the generation of new energy sources and the remediation of environmental pollution.

CO_2 reduction reaction (CO_2RR).

In recent years, the depletion of fossil fuel reserves and the rise in atmospheric CO_2 levels have underscored the urgency of developing sustainable solutions for converting excess CO_2 into valuable chemicals and fuels [12–14]. This not only addresses issues such as the greenhouse effect, the melting of glaciers, and other environmental challenges associated with carbon dioxide, but also provides a potential remedy for the ongoing energy crisis [15]. Carbon dioxide conversion can be achieved through diverse methods, including biochemical [16], electrochemical [17,18], photochemical [19,20], and thermochemical [21] processes. Solar-powered CO_2 reduction, harnessing sunlight as an inexhaustible energy source, has emerged as the most promising approach among these methods [22,23]. Therefore, it has attracted considerable attention, yielding significant advancements [24–27]. Through first-principles calculations, Ju et al. (contribution 1) discovered that introducing selenium vacancies leads to a transition from physical to chemical CO_2 adsorption on Janus WSSe nanotubes. These Se vacancies serve as effective adsorption sites, significantly boosting electron transfer at the interface. This results in heightened electron orbital hybridization between adsorbents and substrates, thereby promising elevated activity and selectivity in CO_2RR. Under illumination, photoexcited holes and electrons generate adequate driving forces, enabling simultaneous oxygen generation reaction (OER) and producing CO_2RR on the sulfur and selenium sides of the defective WSSe nanotube, respectively. This process allows for the reduction of CO_2 into methane, while O_2 is produced through water oxidation, which also supplies hydrogen and electrons for CO_2RR. The findings unveil a potential photocatalyst for achieving efficient photocatalytic CO_2 conversion. Wang et al. (contribution 2) developed an innovative dinuclear Gd (III) complex. This new compound is distinguished by its intricate three-dimensional π–π stacking network, where the voids are occupied by chloride anions and water molecules that are not coordinated. A detailed examination of the complex's Hirschfeld surface indicates the major presence of H$\cdots$H interactions, constituting 48.5% of the surface interactions. These are followed by C$\cdots$H/H$\cdots$C and O$\cdots$H/H$\cdots$O interactions, contributing 27.2% and 6.0%, respectively. The complex showcases notable efficiency in photocatalytic CO_2 reduction experiments, yielding 22.1 µmol/g of CO and 6.0 µmol/g of CH_4 within a span of three hours, with an impressive 78.5% selectivity for CO production. This research sheds light on new possibilities for advancing the study and synthesis of rare earth metal complexes, particularly in the realm of photocatalytic activities for CO_2 reduction. Yuan et al. (contribution 3) delved into the latest progress in solar-powered CO_2 hydrogenation, emphasizing catalyst designs, the architecture of active sites, and the underlying mechanisms. The quest to enhance catalytic efficiency and tackle the hurdles associated with CO_2 reduction has led to the emergence of various innovative strategies. Key among these is the harnessing of light energy, achieved by amplifying the light absorption capacity of catalyst materials. This is primarily facilitated by localized surface plasmon resonances triggered by metal particles (such as Pd, Rh, Ni, Co) and the creation of vacancies. The inclusion of metal particles not only augments optical absorption but also furnishes vital active sites for the activation of H_2 and CO_2 molecules. Furthermore, the introduction of anion vacancies plays a crucial role in surface catalytic reactions by bolstering CO_2 adsorption and diminishing the activation energy required for its reduction. Advanced nanostructures, including photonic crystals, have been meticulously engineered to optimize light utilization. Notably, indium-based oxides, distinguished by their versatile morphologies, phases, and surface-active sites, have demonstrated significant potential in CO_2 hydrogenation. Despite these advancements, several challenges persist, including the limited catalytic activity of numerous catalysts, their prohibitive costs, the notable scarcity of their industrial-scale production, and the difficulties developing specialized apparatus. Moreover, among various photocatalysts, transition metal oxide Cu_2O stands out for its narrow band gap, effective visible light absorption, appropriate conduction band level, affordability, and significant photocatalytic potential. Su et al. (contribution 4) delve into Cu_2O's fundamental characteristics, fabrication techniques, and enhancement approaches. They survey recent Cu_2O-based photocatalysts, as well as their advancements in CO_2 reduction and other applications. The review also points out areas for improvement in Cu_2O-based materials, such as the reliance on costly noble metals for composite and sacrificial agent synthesis, hindering widespread use; challenges in mass-producing high-quality photocatalysts due to potential nanomaterial risks; ongoing issues with photocorrosion that impact durability; and the yet-to-be-clarified catalytic structures and mechanisms of Cu_2O composites.

NH_3 synthesis.

Ammonia, vital in both modern chemical applications and as a structural component in biological molecules, is largely produced industrially via the Haber–Bosch (H-B) method [28]. This method synthesizes ammonia by catalyzing a reaction between nitrogen and hydrogen at elevated temperatures and pressures using metal catalysts [29,30]. Although effective, the H-B method consumes considerable energy and contributes substantially to carbon emissions. To counter this, the "double carbon" initiative promotes combining photocatalytic technologies with synthetic nitrogen fixation, aiming for more sustainable synthetic processes. Researchers harness the sun's abundant energy, using photocatalysts to transform nitrogen/oxynitride into ammonia, a process noteworthy for its energy-saving attributes and simplified storage and transport [31,32]. This approach signifies a shift from solar to chemical energy, paving the way for zero carbon emissions, and thus drastically curbing energy use and environmental impact in industrial ammonia production. Photocatalytic methods in the nitrogen cycle are gaining traction as a significant research field in renewable energy. Zuo et al. (contribution 5) primarily discuss the utilization of composite materials in the field of photocatalytic nitrogen fixation. Their research emphasizes the process of creating NH_3 from N_2 and H_2O using solar power as a renewable energy source under gentle conditions. The study investigates several approaches, including the introduction of defects, the formation of heterojunctions, and the doping of elements, that can be used to adjust the semiconductors' band gap width. These methods aim to enhance the semiconductor's sensitivity to visible light and optimize the usage of light energy. Furthermore, the review introduces the importance of enhancing the movement and separation of photogenerated electrons and holes within the catalysts. This improvement is crucial for increasing the lifespan of these photogenerated carriers and boosting the quantum efficiency of photocatalytic processes. Besides, the photocatalytic NO reduction reaction (NORR) is considered to be a dual-purpose method for both NO removal and NH_3 production. This highlights the necessity for the significant activation of NO molecules by the catalyst, which requires effective chemisorption. Using first-principles calculations, Ju et al. (contribution 6) report a notable transition from physical to chemical adsorption of NO molecules on the Janus WSSe monolayer after introducing Se vacancies. These vacancies potentially serve as optimal adsorption sites, substantially enhancing the electron transfer at the interface. This leads to a pronounced hybridization of electronic orbitals between the adsorbate and the substrate, indicating high NORR activity and selectivity. Furthermore, the spatial constraints imposed by the Se vacancy defects effectively inhibit the N≡N bond coupling and *N diffusion in NO molecules, thus endowing the active site with superior selectivity for NORR in NH_3 synthesis. Additionally, their study reveals that the photocatalytic conversion of NO into NH_3 can spontaneously occur, driven solely by the photo-generated electrons. These insights pave the way for developing highly efficient photocatalysts for NO-to-NH_3 conversion.

Water splitting.

Semiconductor-based photocatalytic water splitting for hydrogen production has gained prominence in sustainable energy research [33–36], especially since the initial discoveries involving TiO_2 photocatalysts [37]. The advent of two-dimensional materials has furthered this field, offering novel photovoltaic and photocatalytic solutions due to their superior properties like extensive specific surface areas, numerous active sites, and reduced carrier migration distances. Huang et al. (contribution 7) examine four δ-IV–VI monolayers, namely, GeS, GeSe, SiS, and SiSe, by employing the first-principles method. These monolayers are noted for their remarkable resilience, with the GeSe monolayer displaying consistent yield strength, even at 30% strain. Notably, the GeSe monolayer boasts an extraordinarily high electron mobility along the x-axis, around 32,507 $cm^2 \cdot V^{-1} \cdot s^{-1}$, significantly surpassing that of its δ-IV–VI counterparts. Additionally, these monolayers' potential in hydrogen evolution reaction capacity, as calculated, suggests promising applications in the realm of photovoltaic and nano-devices.

Degradation of organic dye.

The growth of the textile industry presents notable environmental challenges, especially due to the release of organic dye waste. This wastewater normally contains dyes, such as methyl orange (MO), crystal violet (CV), Congo red (CR), methylene blue (MB), and metanil yellow (MY), or combinations thereof. These dyes, infiltrating water sources for drinking, pose carcinogenic threats to human health [38]. Employing nanomaterials like semiconductors for the photocatalytic breakdown of such dye contaminants offers a potential solution [39]. Wang et al. (contribution 8) utilize a mechanochemical activation-assisted solid-state reaction (MAS) to create $LuFe_{1-x}Co_xO_3$ powders (where x = 0, 0.05, 0.1, 0.15). Investigations using X-ray diffraction (XRD) and Fourier transform infrared (FTIR) spectroscopy reveal that the substitution of B-site iron ions with cobalt ions leads to reductions in lattice parameters. The powder's morphology and elemental composition are further analyzed using scanning electron microscopy (SEM) and energy-dispersive spectroscopy (EDS). UV–visible absorption spectra indicated that $LuFe_{0.85}Co_{0.15}O_3$ powders exhibit a narrower bandgap of 1.75 eV and greater absorbance compared to $LuFeO_3$ (2.06 eV), significantly enhancing light absorption efficiency. Moreover, the $LuFe_{0.85}Co_{0.15}O_3$ powders displayed superior photocatalytic ability over $LuFeO_3$, demonstrating near-complete degradation of methyl orange (MO) in 5.5 h under visible light irradiation with oxalic acid aid. These findings underscore that cobalt doping can notably enhance the photocatalytic efficiency of orthorhombic $LuFeO_3$ for organic dye degradation.

Heterostructures for photocatalysts.

The use of semiconductor-based photocatalysis is garnering significant interest for its potential to harness solar energy directly in order to produce solar-based fuels such as hydrogen and hydrocarbons, in addition to its role in breaking down various pollutants. Despite its promise, the current effectiveness of photocatalytic processes is limited due to rapid recombination of electron–hole pairs and the suboptimal utilization of light. Addressing these challenges has become a focus of substantial research efforts. In particular, the development of specially designed heterojunction photocatalysts has demonstrated enhanced photocatalytic performance, attributable to the effective spatial separation of photogenerated electron–hole pairs [40,41]. Obtaining precision in cultivating or assembling intricate heterostructures remains a considerable challenge. Zhang et al. (contribution 9) explore this issue by examining the collision dynamics of carbon and boron nitride nanotubes under varied collision scenarios, employing self-consistent-charge density-functional tight-binding molecular dynamics. Their study, involving first-principles calculations, reveals the energy stability and electronic structures of heterostructures post-collision. The authors identify five primary collision outcomes: nanotubes can either (1) bounce off each other, (2) join together, (3) merge into a larger, defect-free boron–carbon–nitrogen (BCN) heteronanotube, (4) form a heteronanoribbon combining graphene and hexagonal boron nitride, or (5) suffer severe damage. Notably, both the defect-free (12, 0) BCN heteronanotube and the BCN heteronanoribbon are found to be direct band-gap semiconductors, with band gaps of 0.808 eV and 1.34 eV, respectively. These findings suggest that the electronic structures of nanotubes can be effectively altered through the collision of carbon and boron nitride nanotubes, potentially impacting photocatalytic and other photo-electric applications. While the study initially focuses on the collision fusion of CNTs and BNNTs with specific helicity, its methodology could be applicable to nanotubes of any helicity. Their work not only theoretically introduces a novel method for creating heterostructures via collision fusion but also provides a deeper understanding of the synthesis of heteronanotubes and heteronanoribbons, offering guidance for experimental endeavors. Furthermore, Bissenova et al. (contribution 10) develop a hybrid structure comprising anodic TiO_2 nanotubes intermixed with $SrTiO_3$ particles through chemical synthesis methods. The TiO_2 nanotubes are fabricated by anodization in a solution of ethylene glycol containing NH_4F and H_2O, with a voltage of 30 volts applied. Subsequently, a nanotube array annealed at 450 °C is submerged in a dilute $SrTiO_3$ solution within an autoclave. Scanning electron microscopy (SEM) analysis reveals that the titanium nanotubes are characterized by clear and open ends, boasting an average exterior diameter of 1.01 μm, an interior diameter of 69 nm, and a length measuring 133 nm. Their findings verify the successful creation of a composite structure with potential utility in numerous fields, notably for hydrogen generation through the photocatalytic splitting of water under solar illumination.

It is our sincere hope that the articles published in this Special Issue will contribute to innovation and further in-depth research in the field of photocatalysis. The insights and perspectives presented in each article demonstrate the research results achieved under distinct scenarios. With the impetus of this Special Issue, we eagerly look forward to the emergence of breakthrough research in this field, which will play a key role in addressing the energy crisis and solving the problem of environmental pollution.

Funding: This Special Issue is funded by Young scientist project of Henan province (Grant No. 225200810103), the Program for Science & Technology Innovation Talents in Universities of Henan Province (Grant No. 24HASTIT013), Henan College Key Research Project (Grant No. 24A430002), Natural Science Foundation of Henan Province, (Grant No. 232300420128), Scientific Research Innovation Team Project of Anyang Normal University (Grant No. 2023AYSYKYCXTD04).

Conflicts of Interest: The author declares no conflict of interest.

List of Contributions

1. Ju, L.; Tang, X.; Li, J.; Dong, H.; Yang, S.; Gao, Y.; Liu, W. Armchair Janus WSSe Nanotube Designed with Selenium Vacancy as a Promising Photocatalyst for CO_2 Reduction. *Molecules* **2023**, *28*, 4602. https://doi.org/10.3390/molecules28124602.
2. Wang, L.H.; Tai, X.S. Synthesis, Structural Characterization, Hirschfeld Surface Analysis and Photocatalytic CO_2 Reduction Activity of a New Dinuclear Gd(III) Complex with 6-Phenylpyridine-2-Carboxylic Acid and 1,10-Phenanthroline Ligands. *Molecules* **2023**, *28*, 7595. https://doi.org/10.3390/molecules28227595.
3. Yuan, Z.; Zhu, X.; Jiang, Z. Recent Advances of Constructing Metal/Semiconductor Catalysts Designing for Photocatalytic CO_2 Hydrogenation. *Molecules* **2023**, *28*, 5693. https://doi.org/10.3390/molecules28155693.
4. Su, Q.; Zuo, C.; Liu, M.; Tai, X. A Review on Cu_2O-Based Composites in Photocatalysis: Synthesis, Modification, and Applications. *Molecules* **2023**, *28*, 5576. https://doi.org/10.3390/molecules28145576.
5. Zuo, C.; Su, Q.; Yu, L. Research Progress in Composite Materials for Photocatalytic Nitrogen Fixation. *Molecules* **2023**, *28*, 7277. https://doi.org/10.3390/molecules28217277.
6. Ju, L.; Tang, X.; Zhang, Y.; Li, X.; Cui, X.; Yang, G. Single Selenium Atomic Vacancy Enabled Efficient Visible-Light-Response Photocatalytic NO Reduction to NH_3 on Janus WSSe Monolayer. *Molecules* **2023**, *28*, 2959. https://doi.org/10.3390/molecules28072959.
7. Huang, Z.; Ren, K.; Zheng, R.; Wang, L.; Wang, L. Ultrahigh Carrier Mobility in Two-Dimensional IV-VI Semiconductors for Photocatalytic Water Splitting. *Molecules* **2023**, *28*, 4126. https://doi.org/10.3390/molecules28104126.
8. Wang, Z.; Shi, C.; Li, P.; Wang, W.; Xiao, W.; Sun, T.; Zhang, J. Optical and Photocatalytic Properties of Cobalt-Doped $LuFeO_3$ Powders Prepared by Oxalic Acid Assistance. *Molecules* **2023**, *28*, 5730. https://doi.org/10.3390/molecules28155730.
9. Zhang, C.; Xu, J.; Song, H.; Ren, K.; Yu, Z.G.; Zhang, Y.W. Achieving Boron-Carbon-Nitrogen Heterostructures by Collision Fusion of Carbon Nanotubes and Boron Nitride Nanotubes. *Molecules* **2023**, *28*, 4334. https://doi.org/10.3390/molecules28114334.
10. Bissenova, M.; Umirzakov, A.; Mit, K.; Mereke, A.; Yerubayev, Y.; Serik, A.; Kuspanov, Z. Synthesis and Study of $SrTiO_3/TiO_2$ Hybrid Perovskite Nanotubes by Electrochemical Anodization. *Molecules* **2024**, *29*, 1101. https://doi.org/10.3390/molecules29051101.

References

1. Bie, C.; Wang, L.; Yu, J. Challenges for photocatalytic overall water splitting. *Chem* **2022**, *8*, 1567–1574. [CrossRef]
2. Li, T.; Tsubaki, N.; Jin, Z. S-scheme heterojunction in photocatalytic hydrogen production. *J. Mater. Sci. Technol.* **2024**, *169*, 82–104. [CrossRef]
3. Zhang, J.; Liang, X.; Zhang, C.; Lin, L.; Xing, W.; Yu, Z.; Zhang, G.; Wang, X. Improved Charge Separation in Poly(heptazine-triazine) Imides with Semi-coherent Interfaces for Photocatalytic Hydrogen Evolution. *Angew. Chem. Int. Edit.* **2022**, *61*, e202210849. [CrossRef]
4. Ju, L.; Ma, Y.; Tan, X.; Kou, L. Controllable Electrocatalytic to Photocatalytic Conversion in Ferroelectric Heterostructures. *J. Am. Chem. Soc.* **2023**, *145*, 26393–26402. [CrossRef]
5. Yue, X.; Cheng, L.; Li, F.; Fan, J.; Xiang, Q. Highly Strained Bi-MOF on Bismuth Oxyhalide Support with Tailored Intermediate Adsorption/Desorption Capability for Robust CO_2 Photoreduction. *Angew. Chem. Int. Ed.* **2022**, *61*, e202208414. [CrossRef] [PubMed]
6. Wu, X.; Zhang, W.; Li, J.; Xiang, Q.; Liu, Z.; Liu, B. Identification of the Active Sites on Metallic MoO_{2-x} Nano-Sea-Urchin for Atmospheric CO_2 Photoreduction Under UV, Visible, and Near-Infrared Light Illumination. *Angew. Chem. Int. Ed.* **2023**, *62*, e202213124. [CrossRef] [PubMed]
7. Cheng, L.; Li, B.; Yin, H.; Fan, J.; Xiang, Q. Cu clusters immobilized on Cd-defective cadmium sulfide nano-rods towards photocatalytic CO_2 reduction. *J. Mater. Sci. Technol.* **2022**, *118*, 54–63. [CrossRef]
8. Jing, J.; Yang, J.; Zhang, Z.; Zhu, Y. Supramolecular Zinc Porphyrin Photocatalyst with Strong Reduction Ability and Robust Built-In Electric Field for Highly Efficient Hydrogen Production. *Adv. Energy. Mater.* **2021**, *11*, 2101392. [CrossRef]

9. Ju, L.; Tang, X.; Li, J.; Shi, L.; Yuan, D. Breaking the out-of-plane symmetry of Janus WSSe bilayer with chalcogen substitution for enhanced photocatalytic overall water-splitting. *Appl. Surf. Sci.* **2022**, *574*, 151692. [CrossRef]
10. Ju, L.; Liu, P.; Yang, Y.; Shi, L.; Yang, G.; Sun, L. Tuning the photocatalytic water-splitting performance with the adjustment of diameter in an armchair WSSe nanotube. *J. Energy Chem.* **2021**, *61*, 228–235. [CrossRef]
11. Li, X.; Yu, J.; Jaroniec, M. Hierarchical photocatalysts. *Chem. Soc. Rev.* **2016**, *45*, 2603–2636. [CrossRef] [PubMed]
12. Gattrell, M.; Gupta, N.; Co, A. A review of the aqueous electrochemical reduction of CO_2 to hydrocarbons at copper. *J. Electroanal. Chem.* **2006**, *594*, 1–19. [CrossRef]
13. Gattrell, M.; Gupta, N.; Co, A. Electrochemical reduction of CO_2 to hydrocarbons to store renewable electrical energy and upgrade biogas. *Energy Convers. Manag.* **2007**, *48*, 1255–1265. [CrossRef]
14. Wageh, S.; Al-Hartomy, O.A.; Alotaibi, M.F.; Liu, L.-J. Ionized cocatalyst to promote CO_2 photoreduction activity over core–triple-shell ZnO hollow spheres. *Rare Met.* **2022**, *41*, 1077–1079. [CrossRef]
15. Chen, Y.; Sun, Q.; Jena, P. SiTe monolayers: Si-based analogues of phosphorene. *J. Mater. Chem. C* **2016**, *4*, 6353–6361. [CrossRef]
16. Modestra, J.A.; Mohan, S.V. Microbial electrosynthesis of carboxylic acids through CO_2 reduction with selectively enriched biocatalyst: Microbial dynamics. *J. CO_2 Util.* **2017**, *20*, 190–199. [CrossRef]
17. Cai, F.; Gao, D.; Zhou, H.; Wang, G.; He, T.; Gong, H.; Miao, S.; Yang, F.; Wang, J.; Bao, X. Electrochemical promotion of catalysis over Pd nanoparticles for CO_2 reduction. *Chem. Sci.* **2017**, *8*, 2569–2573. [CrossRef] [PubMed]
18. Wang, Y.-H.; Jiang, W.-J.; Yao, W.; Liu, Z.-L.; Liu, Z.; Yang, Y.; Gao, L.-Z. Advances in electrochemical reduction of carbon dioxide to formate over bismuth-based catalysts. *Rare Met.* **2021**, *40*, 2327–2353. [CrossRef]
19. Qiao, L.; Song, M.; Geng, A.; Yao, S. Polyoxometalate-based high-nuclear cobalt–vanadium–oxo cluster as efficient catalyst for visible light-driven CO_2 reduction. *Chin. Chem. Lett.* **2019**, *30*, 1273–1276. [CrossRef]
20. Wang, Y.; Tian, Y.; Yan, L.; Su, Z. DFT study on sulfur-doped g-C_3N_4 nanosheets as a photocatalyst for CO_2 reduction reaction. *J. Phys. Chem. C* **2018**, *122*, 7712–7719. [CrossRef]
21. Tackett, B.M.; Gomez, E.; Chen, J.G. Net reduction of CO_2 via its thermocatalytic and electrocatalytic transformation reactions in standard and hybrid processes. *Nat. Catal.* **2019**, *2*, 381–386. [CrossRef]
22. Wang, X.-T.; Lin, X.-F.; Yu, D.-S. Metal-containing covalent organic framework: A new type of photo/electrocatalyst. *Rare Met.* **2021**, *41*, 1160–1175. [CrossRef]
23. Zhou, A.-Q.; Yang, J.-M.; Zhu, X.-W.; Zhu, X.-L.; Liu, J.-Y.; Zhong, K.; Chen, H.-X.; Chu, J.-Y.; Du, Y.-S.; Song, Y.-H.; et al. Self-assembly construction of NiCo LDH/ultrathin g-C_3N_4 nanosheets photocatalyst for enhanced CO_2 reduction and charge separation mechanism study. *Rare Met.* **2022**, *41*, 2118–2128. [CrossRef]
24. Muiruri, J.K.; Ye, E.; Zhu, Q.; Loh, X.J.; Li, Z. Recent advance in nanostructured materials innovation towards photocatalytic CO_2 reduction. *Appl. Catal. A-Gen.* **2022**, *648*, 118927. [CrossRef]
25. Luo, Z.; Li, Y.; Guo, F.; Zhang, K.; Liu, K.; Jia, W.; Zhao, Y.; Sun, Y. Carbon Dioxide Conversion with High-Performance Photocatalysts into Methanol on $NiSe_2/WSe_2$. *Energies* **2020**, *13*, 4330. [CrossRef]
26. Biswas, M.; Ali, A.; Cho, K.Y.; Oh, W.C. Novel synthesis of WSe_2-Graphene-TiO_2 ternary nanocomposite via ultrasonic technics for high photocatalytic reduction of CO_2 into CH_3OH. *Ultrason. Sonochem.* **2018**, *42*, 738–746. [CrossRef]
27. Ali, A.; Oh, W.C. Preparation of Nanowire like WSe_2-Graphene Nanocomposite for Photocatalytic Reduction of CO_2 into CH_3OH with the Presence of Sacrificial Agents. *Sci. Rep.* **2017**, *7*, 1867. [CrossRef]
28. Guo, J.; Chen, P. Ammonia history in the making. *Nat. Catal.* **2021**, *4*, 734–735. [CrossRef]
29. Wang, M.; Khan, M.A.; Mohsin, I.; Wicks, J.; Ip, A.H.; Sumon, K.Z.; Dinh, C.-T.; Sargent, E.H.; Gates, I.D.; Kibria, M.G. Can sustainable ammonia synthesis pathways compete with fossil-fuel based Haber–Bosch processes? *Energy Environ. Sci.* **2021**, *14*, 2535–2548. [CrossRef]
30. Smith, C.; Hill, A.K.; Torrente-Murciano, L. Current and future role of Haber–Bosch ammonia in a carbon-free energy landscape. *Energy Environ. Sci.* **2020**, *13*, 331–344. [CrossRef]
31. Khasani; Prasidha, W.; Widyatama, A.; Aziz, M. Energy-saving and environmentally-benign integrated ammonia production system. *Energy* **2021**, *235*, 121400. [CrossRef]
32. Wang, Y.; Meyer, T.J. A Route to Renewable Energy Triggered by the Haber-Bosch Process. *Chem* **2019**, *5*, 496–497. [CrossRef]
33. Chen, S.; Ma, G.; Wang, Q.; Sun, S.; Hisatomi, T.; Higashi, T.; Wang, Z.; Nakabayashi, M.; Shibata, N.; Pan, Z.; et al. Metal selenide photocatalysts for visible-light-driven Z-scheme pure water splitting. *J. Mater. Chem. A* **2019**, *7*, 7415–7422. [CrossRef]
34. Pan, L.; Kim, J.H.; Mayer, M.T.; Son, M.-K.; Ummadisingu, A.; Lee, J.S.; Hagfeldt, A.; Luo, J.; Grätzel, M. Boosting the performance of Cu_2O photocathodes for unassisted solar water splitting devices. *Nat. Catal.* **2018**, *1*, 412–420. [CrossRef]
35. Liu, E.; Jin, C.; Xu, C.; Fan, J.; Hu, X. Facile strategy to fabricate Ni_2P/g-C_3N_4 heterojunction with excellent photocatalytic hydrogen evolution activity. *Int. J. Hydrog. Energy* **2018**, *43*, 21355–21364. [CrossRef]
36. Ju, L.; Shang, J.; Tang, X.; Kou, L. Tunable Photocatalytic Water Splitting by the Ferroelectric Switch in a 2D $AgBiP_2Se_6$ Monolayer. *J. Am. Chem. Soc.* **2020**, *142*, 1492–1500. [CrossRef] [PubMed]
37. Fujishima, A.; Honda, K. Electrochemical Photolysis of Water at a Semiconductor Electrode. *Nature* **1972**, *238*, 37–38. [CrossRef]
38. Chen, S.; Zhang, J.; Zhang, C.; Yue, Q.; Li, Y.; Li, C. Equilibrium and kinetic studies of methyl orange and methyl violet adsorption on activated carbon derived from Phragmites australis. *Desalination* **2010**, *252*, 149–156. [CrossRef]
39. Lanjwani, M.F.; Tuzen, M.; Khuhawar, M.Y.; Saleh, T.A. Trends in photocatalytic degradation of organic dye pollutants using nanoparticles: A review. *Inorg. Chem. Commun.* **2024**, *159*, 111613. [CrossRef]

40. Ju, L.; Dai, Y.; Wei, W.; Li, M.; Huang, B. DFT investigation on two-dimensional GeS/WS 2 van der Waals heterostructure for direct Z-scheme photocatalytic overall water splitting. *Appl. Surf. Sci.* **2018**, *434*, 365–374. [CrossRef]
41. Low, J.; Yu, J.; Jaroniec, M.; Wageh, S.; Al-Ghamdi, A.A. Heterojunction Photocatalysts. *Adv. Mater.* **2017**, *29*, 1601694. [CrossRef] [PubMed]

Disclaimer/Publisher's Note: The statements, opinions and data contained in all publications are solely those of the individual author(s) and contributor(s) and not of MDPI and/or the editor(s). MDPI and/or the editor(s) disclaim responsibility for any injury to people or property resulting from any ideas, methods, instructions or products referred to in the content.

Review

Research Progress in Composite Materials for Photocatalytic Nitrogen Fixation

Cheng Zuo, Qian Su and Lei Yu *

College of Chemistry & Chemical and Environmental Engineering, Weifang University, Weifang 261061, China; 17854270427@163.com (C.Z.); sqian316@wfu.edu.cn (Q.S.)
* Correspondence: jimoyulei@163.com

Abstract: Ammonia is an essential component of modern chemical products and the building unit of natural life molecules. The Haber–Bosch (H-B) process is mainly used in the ammonia synthesis process in the industry. In this process, nitrogen and hydrogen react to produce ammonia with metal catalysts under high temperatures and pressure. However, the H-B process consumes a lot of energy and simultaneously emits greenhouse gases. In the "double carbon" effect, to promote the combination of photocatalytic technology and artificial nitrogen fixation, the development of green synthetic reactions has been widely discussed. Using an inexhaustible supply of sunlight as a power source, researchers have used photocatalysts to reduce nitrogen to ammonia, which is energy-dense and easy to store and transport. This process completes the conversion from light energy to chemical energy. At the same time, it achieves zero carbon emissions, reducing energy consumption and environmental pollution in industrial ammonia synthesis from the source. The application of photocatalytic technology in the nitrogen cycle has become one of the research hotspots in the new energy field. This article provides a classification of and an introduction to nitrogen-fixing photocatalysts reported in recent years and prospects the future development trends in this field.

Keywords: nitrogen; ammonia synthesis; photocatalysis technology; photocatalysts

Citation: Zuo, C.; Su, Q.; Yu, L. Research Progress in Composite Materials for Photocatalytic Nitrogen Fixation. *Molecules* **2023**, *28*, 7277. https://doi.org/10.3390/molecules28217277

Academic Editors: Lin Ju and Stefano Falcinelli

Received: 10 September 2023
Revised: 17 October 2023
Accepted: 24 October 2023
Published: 26 October 2023

Copyright: © 2023 by the authors. Licensee MDPI, Basel, Switzerland. This article is an open access article distributed under the terms and conditions of the Creative Commons Attribution (CC BY) license (https://creativecommons.org/licenses/by/4.0/).

1. Introduction

With the rapid development of the global economy, energy sources and the environment are being irreversibly damaged, threatening the survival and development of humankind. It is urgent to find solutions to the energy crisis and environmental pollution. Ammonia is one of the most highly produced chemicals in the world [1]. The progress of production directly affects the energy structure and environmental issues. Currently, global NH_3 production is approximately 170 million tons per year and highly relies on the traditional Haber–Bosch (H-B) process [2,3]. Industrial ammonia synthesis is usually carried out at high temperatures to improve the reaction rate and maintain optimal catalyst activity (Table 1). At the same time, the H-B process uses high pressure to overcome thermodynamic limitations and promote a rightward shift in reaction equilibrium, thereby improving conversion rates. Despite the harsh reaction conditions, the one-way conversion rate of synthetic NH_3 can only reach 10–15%. In addition, the H-B process is powered by energy from the reforming or gasification of natural gas and fossil fuels using pressurized superheated steam. The H_2 production process consumes about 75% of the energy input and produces half as much carbon dioxide as the entire process. The annual emissions of CO_2 greenhouse gases from the entire H-B process amount to 300 million tons, accounting for approximately 1.6% of the global total emissions (Figure 1a) [4]. Therefore, finding a new substance to replace H_2 as a proton source while overcoming harsh reaction conditions is an ideal method to reduce fossil energy consumption and CO_2 emissions [5]. Photocatalytic nitrogen fixation technology utilizes renewable solar energy as the energy source to achieve the catalytic synthesis of NH_3 from N_2 and water under mild conditions. In addition, photocatalysis, capable of promoting thermodynamic non-spontaneous N_2 reduction reactions, would be a green and sustainable alternative to the H-B process by continuously supplying electrons to activate adsorbed molecules to reduce N_2 in synthesizing NH_3 (Figure 1b).

Table 1. Comparison between the Haber–Bosh process and photocatalytic nitrogen fixation process.

	Haber–Bosh	Photocatalytic Nitrogen Fixation
Reaction equation	$N_2 + 3H_2 \rightarrow 2NH_3$	$2N_2 + 6H_2O \rightarrow 4NH_3 + 3O_2$
Hydrogen source	Natural gas	Water
Catalysts	Fe/Ru-based catalysts	Mainly semiconductors
Temperature	400–600 °C	Room temperature
Pressure	150–300 atm	1 atm
Energy source	Fossil fuel	Solar energy

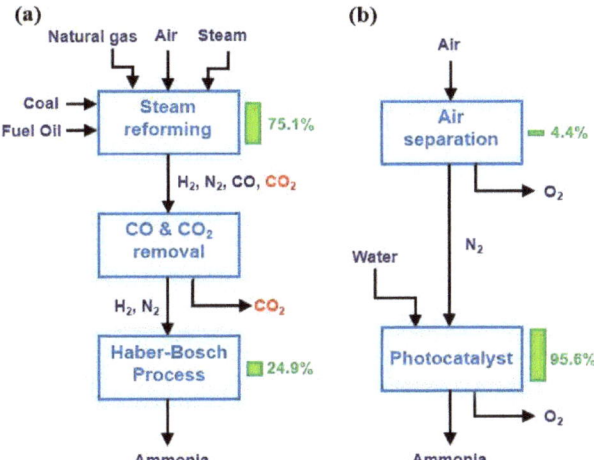

Figure 1. Energy efficiency analysis of (a) the H-B process and (b) photocatalytic nitrogen fixation for NH_3 synthesis. The columns and numbers to the right of the block represent the share of the total energy input [4]. Copyright 2018, Cell Press.

In the nitrogen reduction reaction, the efficiency of photocatalytic ammonia synthesis remains low due to the inert structure of the N_2 molecule, the difficulty in dissociating the N≡N bond, and the presence of high-energy intermediates (e.g., N_2H) [6,7]. The low reaction efficiency severely limits the development and application of photocatalysis. Enhancing the reaction activity using efficient photocatalysts is the core of photocatalytic nitrogen fixation. In addition, the insolubility of nitrogen limits the gas–liquid contact of the non-homogeneous reaction system, which reduces the efficiency of the photocatalytic reaction. Therefore, establishing a new photocatalytic reaction system to improve the utilization of visible light and enhance the gas–liquid mass transfer ability is an essential means to realize the high efficiency and stability of photocatalytic nitrogen fixation reactions.

Currently, the research on photocatalysts for reducing nitrogen is mainly focused on improving the overall reaction efficiency, including light absorption, the separation and migration of photogenerated carriers, and the surface-catalyzed reaction. Several common photocatalytic materials, such as metal oxides, metal sulfides, bismuth halides, carbon-based materials, and MOFs, are presented in this paper. The semiconductor materials' morphology, structure, and properties are analyzed to construct tunable catalytic systems. A structural morphology with a large specific surface area is prepared by changing the reaction conditions, precursor composition and ratio, and using other strategies to improve the contact chances between the catalyst and the reactants, increase the active sites, promote the adsorption and activation of N_2, the rapid dissociation of the N≡N bond, as well as the reduction in the production of energetic intermediates. In addition, the semiconductor bandgap width was adjusted by introducing vacancies, constructing heterojunctions, and elemental doping to reduce the bandgap, realize the catalyst's response to visible light, and improve light energy utilization [8–10].

In future research, we will analyze the mechanism of photocatalysts and reactions in combination with the density functional theory (DFT) and the feedback of the experimental results. We strive to create a set of reasonable and efficient photocatalytic reaction systems tailored for nitrogen fixation to achieve high energy utilization and excellent catalytic activity in the nitrogen fixation process.

2. Photocatalysts for Nitrogen Fixation

The photocatalyst is one of the critical factors determining whether the photocatalytic nitrogen fixation process can be carried out smoothly. So far, researchers have conducted systematic studies on designing efficient photocatalytic materials and developed numerous photocatalysts that could realize nitrogen reduction under mild conditions. Depending on the elemental composition, the materials used for nitrogen fixation reactions include metal oxides, metal sulfides, bismuth halides, carbon-based materials, and MOFs.

2.1. Metal Oxides

The use of metal oxides as materials for photocatalytic nitrogen fixation dates back to the pioneering work of Schrauzer and Guth in 1977 [11]. Their study successfully reduced nitrogen to ammonia and a small amount of hydrazine (N_2H_4) using Fe_2O_3-doped TiO_2 as the catalyst and UV light and water as the light and proton sources, respectively. By adjusting the Fe doping amount, the experiment showed that TiO_2 containing 0.2% Fe had the best ammonia production rate. Inspired by this, Augugliaro et al. [12] prepared a series of Fe-doped TiO_2 using co-precipitation and impregnation techniques to investigate the nitrogen fixation activity of the samples in a continuous fixed-bed reactor, to analyze the roles of OH and Fe^{3+} on the surface of the catalysts in the reaction process, and to probe the reaction nature of photocatalysis. Radford et al. [13] synthesized Fe-doped anatase and rutile TiO_2 by metal vaporization. It was found that the undoped samples could not drive the nitrogen reduction reaction, whereas the Fe-doped samples were endowed with catalytic activity to drive the reaction, and the Fe-doped anatase had more negative flat-band potential energy, thus having higher activity. Based on in-depth investigation, the mechanism of Fe in the photocatalytic reaction was mainly reflected in two aspects: On the one hand, the appropriate amount of Fe doping could capture photogenerated electrons and inhibit the recombination of photogenerated carriers [14]. On the other hand, Fe doping could generate oxygen vacancies and corresponding defect energy levels, and the high spin state Fe(III) prompts Fe 3d electrons to feedback to the N $1\pi g^*$ orbitals to activate the adsorbed nitrogen molecules [15]. Other transition metals such as Ru, Co, Mo, and Ni have been shown to contribute to the catalytic performance when introduced as dopants into TiO_2 [16,17]. In addition to element doping, constructing heterostructures to improve the separation and transportation of photogenerated charges is also an effective means to enhance photocatalytic activity and stability. For example, TiO_2/Cu_7S_4 composites were loaded onto copper mesh by hydrothermal and calcination methods, forming an S-scheme heterojunction at the interface [18]. The calcination treatment increases the specific surface area and surface defects of the photocatalyst. The rich oxygen vacancies and S-scheme heterostructures of photocatalysts accelerate the separation and transport of photogenerated carriers, resulting in a strong redox ability of photocatalysts. Under visible light, the yield of NH_3 synthesized by the OV-TiO_2@Cu_7S_4 photocatalyst reached 133.42 $\mu mol \cdot cm^{-2} \cdot h^{-1}$, which is 5.2 and 2.2 times that of pure TiO_2 and Cu_7S_4, respectively.

In addition to TiO_2, other metal oxides such as iron oxide (Fe_2O_3) [19,20], bismuth oxide (BiO) [21], tungsten oxide (WO_3) [22,23], zinc oxide (ZnO) [24,25], and gallium oxide (Ga_2O_3) [26,27] have been used as candidates for photocatalytic nitrogen fixation materials. Khader et al. [19] used α-Fe_2O_3 partially reduced to Fe_3O_4 in the presence of 3–5% divalent iron ions in the catalyst, and ammonia production was detected in the catalyst slurry by UV irradiation. Fe_2O_3 was shown to be an effective photocatalyst for nitrogen reduction, and its narrow bandgap feature enabled response to visible light [28]. Wang et al. [11] used a simple hydrothermal synthesis method to prepare low-valent Bi^{2+} containing BiO materials for photocatalytic nitrogen fixation. As shown in Figure 1a, unlike ordinary Bi^{3+}, Bi^{2+} in BiO has empty 6d orbitals that accept electrons from N_2 and provide high-quality chemisorption and activation centers. N_2 was activated by three aligned Bi atoms by supplying electrons to the 6d orbitals of Bi and accepting lone pairs of electrons from the three Bi atoms into their empty antibonding orbitals (σ^*2p_x, π^*2p_y, and π^*2p_z), generating a $1N_2$-3Bi(II) side-pair bonding structure, which significantly weakened the N≡N bond and accelerated the photocatalytic NRR process. Hao et al. [29] employed nanostructured Bi_2MoO_6 crystals as a novel photocatalyst for synthesizing ammonia from air and water molecules without adding any sacrificial agent. The significantly improved photocatalytic nitrogen fixation performance (1.3 mmol·g_{cat}^{-1}·h^{-1}) was mainly attributed to the ligand-unsaturated Mo atoms exposed at the edges of the MoO_6 polyhedra becoming the active centers to promote the chemisorption activation process of N_2. Introducing oxygen vacancies or noble metals on the surface to construct active centers was the key to improving photocatalytic activity for the WO_3 and ZnO. According to Hou et al. [22], the grain boundaries (GBs) in nanoporous WO_3 were induced to produce abundant surface defects under light, which were able to modulate the energy band structure, enhance the W-O covalency, and drive the photogenerated electron transfer to adsorbed N_2. This significantly enhanced the nitrogen-fixing activity of WO_3-600. Janet et al. [24] used wet etching and chemical precipitation to synthesize Pt-loaded ZnO with increased active centers resulting in a reactive ammonia yield of 86 μmol·g_{cat}^{-1}·h^{-1} at ambient temperature and pressure (Figure 2a). Zhao et al. [26] used uniformly stabilized mesoporous β-Ga_2O_3 nanorods as photocatalysts for photocatalytic nitrogen fixation under UV light irradiation (λ = 254 nm). The broad bandgap of the synthesized β-Ga_2O_3 material was about 4.4 eV, which effectively suppressed the complexation of photogenerated carriers, and a quantum yield of up to 36.1% for nitrogen fixation was obtained by the combined effect of in situ-grown CO_2-induced electron transfer and photocatalyst surface electron transfer (Figure 2b). Meanwhile, methanol, ethanol, n-propanol, and n-butanol were employed as hole-trapping agents to further improve the conversion efficiency.

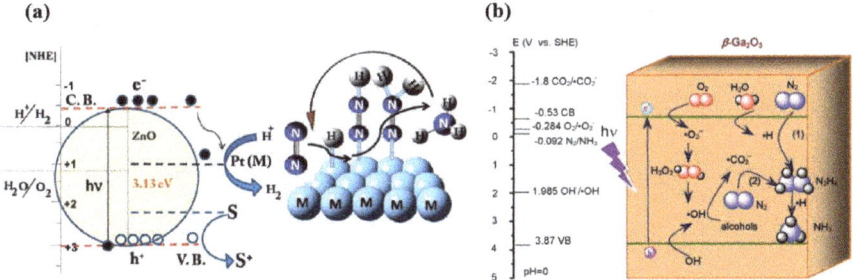

Figure 2. (a) Schematic representation of simultaneous hydrogen production and nitrogen reduction on Pt-doped ZnO [24]. (b) Possible direct and indirect electron transfer pathways on β-Ga_2O_3 photocatalysts [26]. Copyright 2017, Elsevier.

So far, metal oxides (mainly non-precious metal oxides) have attracted much attention because of their advantages such as easy synthesis, stability and control, low cost, and environmental friendliness. For example, $SrTiO_3$, which is widely used in the field of water cracking, has also received some attention in photocatalytic ammonia synthesis. However, there is no universal consensus on the mechanism of photocatalytic reduction of N_2 by metal oxides. Based on theoretical calculation, the dissociative mechanism and associative mechanism for nitrogen fixation have been gradually explored and tested. In recent years, some research results have provided new ideas and prospects for the application of metal oxides in photocatalytic NRR.

2.2. Metal Sulfides

Metal sulfides have excellent optical, electrical, and magnetic properties, and their narrow bandgap facilitates the absorption of visible light to obtain high light energy utilization. Khan et al. [30] used CdS/Pt/RuO$_2$ composite to reduce N$_2$ under visible light (λ = 505 nm) irradiation, and the activated dinitrogen reacted with [Ru(Hedta)(H$_2$O)]$^-$ to produce [Ru(Hedta)(N$_2$)]$^-$ complex. A continuous supply of photogenerated electrons from CdS to this complex reacts to form ammonia. As the photoreaction proceeded, the ammonia yield decreased due to photocorrosion by CdS. To improve the photocatalytic activity and stability, Ye et al. [31] used a Cd$_{0.5}$Zn$_{0.5}$S solid solution for photocatalytic nitrogen fixation for the first time and employed a transition metal phosphide (Ni$_2$P) as a co-catalyst. Ni$_2$P/Cd$_{0.5}$Zn$_{0.5}$S was used for photocatalytic nitrogen reduction reaction without adding any sacrificial agent. After irradiation with visible light (λ > 400 nm) for 1 h, the NH$_3$ concentration reached 101.5 μmol·L^{-1}. The quantum efficiency under 420 nm monochromatic light reached 4.32%, much higher than those of other semiconductors. As tested by time-resolved fluorescence spectroscopy, photocurrent, and electrochemical impedance spectroscopy, the samples with the addition of the co-catalysts rapidly transferred the photogenerated electrons to Ni$_2$P through excellent heterogeneous interfacial contacts to reduce the charge complexation, thus improving the photogenerated carrier separation efficiency (Figure 3a). In addition, the photogenerated electron–hole pairs in the ultrathin transition metal sulfides (TMDs) could form tightly bound excitons, which give very high dissociation energies by trapping electrons. As a member of TMDs, MoS$_2$ is getting much attention [32–34]. Sun et al. [35] found that ultrasonically treated ultrathin MoS$_2$ could photocatalytically reduce nitrogen to ammonia with a photocatalytic ammonia yield of up to 325 μmol·g$_{cat}^{-1}$·h^{-1} without the use of a sacrificial agent or co-catalyst, and had considerable stability. Photogenerated excitons captured the free electrons in the ultrathin MoS$_2$ to generate charged excitons near the Mo sites, which interacted with the adsorbed N$_2$ to promote the multi-electron transfer, lower the reaction thermodynamic potential barrier, and accelerate the process of the photocatalytic reduction of nitrogen (Figure 3b).

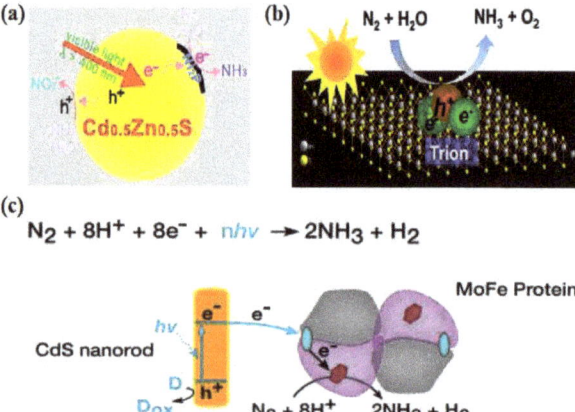

Figure 3. (a) Photocatalytic nitrogen fixation mechanism of Ni$_2$P/Cd$_{0.5}$Zn$_{0.5}$S [31]; copyright 2017, Elsevier. (b) Schematic diagram of the exciton-induced multi-electron N$_2$ reduction process [35]; copyright 2017, Elsevier. (c) Catalytic reaction mechanism of CdS: MoFe protein complexes [36]; copyright 2016, Elsevier.

Inspired by nitrogen-fixing enzymes, researchers have studied photocatalytic nitrogen reduction reactions in the cross-fertilized materials science and biology disciplines. Brown et al. [36] adsorbed MoFe proteins (the active site of nitrogen-fixing enzymes) onto CdS nanorods to form biological nanocomplexes and investigated their nitrogen-fixing activities. Photosensitization of the MoFe protein using CdS nanocrystals replaced ATP hydrolysis by capturing light energy (Figure 3c). Under visible light, the ammonia production rate reached 315 $\mu mol \cdot mg^{-1} \cdot min^{-1}$, which was on par with the biological nitrogen-fixing enzyme capacity. Given the prominent role of the MoFe factor in nitrogen-fixing enzymes, Banerjee et al. [37] deduced that solid compounds consisting of FeMoS inorganic clusters could reduce nitrogen in water to ammonia in the presence of light, and thus a combination of $[Mo_2Fe_6S_8(SPh)_3]^{3+}$ and $[Sn_2S_6]^{4-}$ clusters was used to constitute bionic sulfur compounds. The designed and synthesized $Fe_2Mo_6S_8$ thiocolloid has strong light absorption, high specific surface area, and excellent water stability. Thus, its performance was superior to that of nitrogen-fixing enzymes. On this basis, Liu et al. [38] designed a novel thioglycolic system consisting of $Fe_2Mo_6S_8(SPh)_3$ and Fe_3S_4 mimetic clusters. The bonding between nitrogen and iron was determined using local orbital theory analysis, demonstrating that Fe was the active site for N_2 binding and that it drives the nitrogen reduction reaction more readily than the Mo metal site [39,40].

Metal-sulfide-based photocatalysts have relatively narrow band gaps, abundant active sites, and adjustable electronic properties, which are suitable for nitrogen fixation. However, the metal sulfides applied to the photocatalytic reduction of N_2 to date are mainly based on CdS, and the ammonia production rate is generally low. For other metal sulfides, such as two-dimensional metal disulfide and indium-based sulfide, the potential of catalytic nitrogen fixation has been preliminarily predicted by theory and experiment. Considering the diversity of metal sulfides, such catalysts need to be further explored as efficient artificial nitrogen fixation catalytic materials.

2.3. BiOX-Based Materials

Bismuth halide oxide (BiOX, X = Cl, Br, I) has attracted much attention due to its superior optical properties. Its layered structure provides ample space for atomic polarization and an internal electric field that facilitates the separation and transfer of photogenerated carriers [41–43]. The application of BiOX-based materials in photocatalytic nitrogen fixation has been demonstrated in recent works.

Li et al. [44] demonstrated that the photocatalytic reduction reaction of nitrogen could be realized under visible light without any organic sacrificial agent or precious metal co-catalyst using BiOBr nanosheets at room temperature and pressure. The prepared catalysts possessed electron-donating properties upon photoexcitation, driving the interfacial electron transfer from BiOBr nanosheets to adsorbed N_2, and ammonia yields as high as 104.2 $\mu mol \cdot g_{cat}^{-1} \cdot h^{-1}$ were obtained. Combined with theoretical simulations, the oxygen vacancies in BiOBr extend the activated N≡N bond length from 1.078 Å to 1.133 Å, promoting the activation of nitrogen molecules. Due to the generation of abundant oxygen vacancies on the surface, a defect state was formed at the bottom of the BiOBr conduction band, which inhibits the recombination of electron–hole pairs. In addition, the group examined the photocatalytic activity of BiOCl containing abundant oxygen vacancies [45]. The kinetics and mechanisms of the photocatalytic reactions differed due to the different exposed crystalline surfaces. The mechanism of nitrogen fixation on the (110) crystalline face follows a distal binding mechanism ($N_2 \rightarrow \bullet N\text{-}NH_3 \rightarrow \bullet N + NH_3 \rightarrow 2NH_3$), while the reaction on the (010) face follows an alternating binding mechanism ($N_2 \rightarrow N_2H_3 \rightarrow N_2H_4$). Under UV irradiation at a wavelength of 254 nm, the quantum yields of the BiOCl (001) and (010) crystal faces were 1.8% and 4.3%, respectively. To further demonstrate the effect of exposed crystal faces on photocatalytic activity, Bai et al. [46] prepared Bi_5O_7I nanosheets with different exposed crystal faces, in which the nitrogen fixation activities of the catalyst samples with exposed crystal faces of (001) and (100) were 111.5 $mmol \cdot L^{-1} \cdot h^{-1}$ and 47.6 $mmol \cdot L^{-1} \cdot h^{-1}$, respectively. The difference was due to the higher photogenerated carrier separation efficiency and more negative conduction band position (−1.45 eV) in Bi_5O_7I-001. Zeng et al. [47] successfully synthesized carbon-doped BiOI (C-BiOI) by hydrothermal reaction, demonstrating that the surface carbon elements adsorb nitrogen. The ammonia yield of C-BiOI-3 under visible light was as high as 311 $\mu mol \cdot g_{cat}^{-1} \cdot h^{-1}$, about 3.7 times higher than that of pure BiOI. Carbon clusters entered the intercalation of BiOI crystals during the preparation process, interfered with the periodicity of the crystal lattice, and induced the generation of vacancies in the BiOI structure, which resulted in a decrease in the catalyst band gap and enhancement in visible light absorption, and the trapping of photogenerated electrons by the vacancies, which led to improvement in the charge separation efficiency and accelerated the photocatalytic reaction. In addition, carbon doping affected the morphology of the catalysts with reduced crystal size and increased specific surface area, facilitating the contact between the catalysts and reactants. However, the induced surface oxygen vacancies in BiOX-based materials were easily oxidized during the reaction process, decreasing photocatalytic NRR activity. To alleviate this difficulty, Wang et al. [48] designed ultrafine Bi_5O_7Br nanotubes with abundant sustainable oxygen vacancies to accelerate the photocatalytic reduction of nitrogen in aqueous solvent in order to synthesize ammonia without the addition of any sacrificial agents or co-catalysts. The synthesized sample has a large specific surface area (>96 $m^2 \cdot g^{-1}$), suitable light-absorbing band edges, and a continuous supply of surface oxygen vacancies, and thus the ammonia yield obtained is as high as 1.38 $mmol \cdot g_{cat}^{-1} \cdot h^{-1}$, and the apparent quantum efficiency at 420 nm is close to 2.3%.

The indirect bandgap of BiOX material effectively hinders charge recombination. Its unique layered structure not only facilitates the generation of vacancies as active sites for catalytic reactions but also provides internal electric fields as driving forces for charge transfer. In addition, research was conducted on the photocatalytic reduction of N_2 using BiOX substrate materials from the perspectives of defect engineering, surface engineering, and band gap structure adjustment. It is worth noting that high-quality 2D BiOX-based materials have a photocatalytic surface that changes with the progress of photo reactions and can serve as a dynamic crystal model for theoretical simulation. The combination of a dynamic simulation algorithm and experimental data can be used as a new simulation method to deeply understand the photocatalytic reaction mechanism.

2.4. Carbon-Based Materials

Carbon-based materials commonly used for photocatalysis include diamond, graphene, carbon nanotubes, and graphitic carbon nitride. Zhu et al. [49] prepared boron-doped diamonds to catalyze ammonia synthesis by nitrogen reduction under mild conditions. Transient absorption tests at a wavelength of 632 nm showed that diamond transfers solvated electrons to water when photoexcited. Comparative tests using samples and purchased product powders showed that the photocatalytic activity depended on the H terminals on the diamond surface and was correlated with the production of solvated electrons. In this catalytic process, the electrons were transported directly to the reactants without going through molecular adsorption on the catalyst's surface, making it a new paradigm for photocatalytic reduction. Diamond's stability and acid resistance set it apart from conventional photovoltaic materials. Bandy et al. [50] synthesized diamond thin films on Mo, Ni, and Ti metal substrates, and photoresponse tests showed that H-terminated thin films with a negative electron affinity drove nitrogen reduction.

In contrast, O-terminated thin films showed almost no photocatalytic activity. The electrons in the metal substrate were transferred to the conduction band of the diamond through a barrier-free electron emission process, thus providing enough energy to participate in the nitrogen fixation reaction (Figure 4a). Graphene, as an allotrope of diamond with excellent electrical conductivity, is also considered an excellent substrate with the ability to activate N_2. Tian et al. [51] demonstrated the ability of aluminum-doped graphene to convert nitrogen to ammonia through DFT simulations. Li et al. [52] proposed that FeN_3-embedded graphene could be used as a raw material for photocatalytic nitrogen reduction through first-principle calculations. In addition, Perathoner et al. [53] used carbon nanotubes loaded with Fe as the photocatalyst to harvest an ammonia yield of 2.2×10^{-3} g·m^{-2}·h^{-2} at ambient temperature and pressure. Liu et al. [54] prepared nitrogen-doped porous carbon (NPC) using pyrolysis of an imidazolium zeolite skeleton, which is a structure with high N content and tunable N species, to promote nitrogen molecule chemisorption and activation, thus addressing the problem of the slow kinetics of nitrogen fixation reactions.

The lack of active sites and photogenerated carriers in pure carbon materials limits their nitrogen fixation applications. Therefore, researchers have developed graphitic carbon nitride (g-C$_3$N$_4$)-based photocatalysts. Dong et al. [55] successfully synthesized g-C$_3$N$_4$ containing nitrogen vacancies by nitrogen heat treatment and reported the effect of nitrogen vacancies on the activity of semiconductor photocatalytic nitrogen reduction reactions. In the photocatalytic experiments, it was observed that the nitrogen vacancies endowed the g-C$_3$N$_4$ with photocatalytic nitrogen fixation ability. Since nitrogen vacancies have the same shape and size as nitrogen atoms, they could selectively adsorb activated nitrogen, and thus the photocatalytic nitrogen fixation process did not interfere with other gases. In addition to this advantage, nitrogen vacancies improve the separation efficiency of photogenerated carriers and promote the transfer of photogenerated electrons from g-C$_3$N$_4$ to adsorbed N$_2$. Wu et al. [56] prepared a spongy g-C$_3$N$_4$, whose excellent nitrogen fixation capability benefited from the trapping of photogenerated electrons by the surface nitrogen vacancies (Figure 4b). Cao et al. [57] used urea as the raw material and, using a simple one-step separation method, synthesized amine-functionalized ultrathin g-C$_3$N$_4$ nanosheets. Compared with bulk g-C$_3$N$_4$, the synthesized g-C$_3$N$_4$ nanosheets have a larger specific surface area, higher reduction potential and carrier separation efficiency, and improved photocatalytic activity and stability of nitrogen fixation reaction under visible light irradiation.

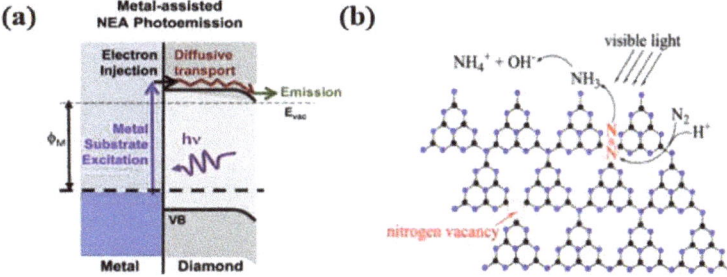

Figure 4. (a) Metal-assisted NEA photoelectron emission process [50]; copyright 2016, Elsevier. (b) Possible photocatalytic nitrogen fixation on M-GCN [56]; copyright 2016, Elsevier.

Li et al. [58] pretreated the samples with alkali solutions of appropriate concentrations, which resulted in the C=N bond breaking and surface K$^+$ grafting of the homotriazine structural unit in g-C$_3$N$_4$. The g-C$_3$N$_4$ etched with KOH solution was used as the photocatalytic material, while methanol solution was introduced as the proton source for the first time. The photocatalyst showed an ammonia yield of 3.632 mmol·g^{-1}·h^{-1} and an apparent quantum efficiency of up to 21.5% under light irradiation with a wavelength of ~420 nm. The enhancement in the catalytic activity was due to the combined effects of various aspects, including the enhanced absorption of light, the fast dissolution of N$_2$ in methanol solution, the increase in active centers on the catalyst surface, and the promotion of carrier transfer and ammonia desorption by CH$_3$OH and surface K$^+$. In addition to structural optimization, strategies such as elemental doping and material composites have also been used as modification methods for g-C$_3$N$_4$. For example, Hu et al. [59] synthesized honeycomb-shaped Fe-doped g-C$_3$N$_4$ by controlling the concentration of Fe^{3+}. Fe^{3+} enters the lattice sites and stably exists in the structure of g-C$_3$N$_4$ through Fe-N coordination bonding. The Fe metal sites became the active centers for the adsorption and activation of nitrogen molecules and promoted the interfacial charge transfer between the catalyst and the nitrogen molecules, significantly improving nitrogen fixation capacity. The best ammonia yield was obtained for Fe0.05-CN, about 5.40 mg·L^{-1}·h^{-1}·g$_{cat}$$^{-1}$, close to 13.5 times that of pure g-C$_3$N$_4$. In addition, the photocatalytic nitrogen fixation performance of g-C$_3$N$_4$ semiconductor-based composites, such as MnO$_{2-x}$/g-C$_3$N$_4$, Ti$_3$C$_2$/g-C$_3$N$_4$, and g-C$_3$N$_4$/FeOCl, was significantly improved due to the construction of heterostructures to promote the separation and transfer of photogenerated carriers [60,61].

At present, carbon-based materials applied in the photocatalytic reduction of N$_2$ are mainly g-C$_3$N$_4$ and 2D graphene. Due to their unique structure, excellent light absorption performance, and conductivity, both are considered nitrogen-fixing photocatalysts with infinite potential. In addition, inspired by the size-dependent photoluminescence effect, modification of carbon-based materials by morphological regulation is an effective way to improve the photoactivity of catalysts. For example, the design of zero-dimensional carbon quantum dots enables carbon-based materials to obtain the advantages of adjustable chemical structure, high quantum efficiency, and good biocompatibility, so as to improve catalyst reaction activity.

2.5. MOFs and Derivatives

Metal-organic skeletons (MOFs), as porous materials with high crystallinity and tunable organic ligands, have been shown to have efficient photocatalytic activity. In recent years, researchers have designed a series of MOFs and their derivatives and performed photocatalytic nitrogen reduction experiments [62–65].

Huang et al. [66] prepared NH$_2$-MIL-125(Ti) catalysts by integrating metal sites with amine-based functional groups and applied them to photocatalytic nitrogen fixation reaction at ambient temperature and pressure, obtaining an ammonia yield of 12.3 μmol·g^{-1}·h^{-1} under visible light. Through ligand functionalization, the light absorption range of the MOF materials was extended to the visible light region. Simultaneous electron transfer between the ligand and the metal-induced Ti^{3+} production provides abundant active sites for nitrogen reduction (Figure 5a). Inspired by chlorophyll, Shang et al. [67] developed a porphyrin-based metal-organic skeleton (PMOF) with Fe as the active center, with Al characterized as the metal node with excellent stability, and Fe atoms dispersed on each porphyrin ring to facilitate nitrogen adsorption activation. Calculations showed that the Fe-N site in Al-PMOF(Fe) acts as the active center of the photocatalytic reaction and reduces the difficulty of the rate-determining step in the reaction process (Figure 5b). The ammonia yield of Al-PMOF(Fe) was 127 μg·g^{-1}·h^{-1}, which was a 50% improvement in performance compared to the pristine Al-PMOF catalysts. Zhang et al. [68] simulated the π-orbitals of the reverse feeding mechanism of a designed and synthesized MOF-76(Ce) material, in which Ce acts as the active center for capturing photogenerated electrons. Experimental results and theoretical analyses showed that the presence of cerium metal in a ligand-unsaturated state (Ce-CUS) on the surface of MOF-76(Ce) nanorods could provide unoccupied 4f orbitals to collect electrons and transfer them to N$_2$. The synthesized materials exhibited excellent photocatalytic nitrogen reduction performance with ammonia yields as high as 34 μmol·g^{-1}·h^{-1} at ambient temperature and pressure. Xu et al. [69] successfully synthesized CeZr$_5$-MOF(UiO-66) bimetallic photocatalysts using the rare earth element Ce to partially replace Zr. Ce was introduced into the nodes of Zr-MOFs(UiO-66) to form CeZr$_5$ clusters, which enhanced the separation and transfer rate of the photogenerated electron–hole pairs through the charge-transfer process between the ligand and the metal, thus enhancing the photocatalytic nitrogen fixation activity. In addition, the photocatalytic performance was increased linearly with the increase in Ce content when the Ce content was lower than 20%. The photocatalytic nitrogen fixation activity was 200.13 μmol·g^{-1}·h^{-1}, 105.9% higher than that of Zr-UiO-66. Zhao et al. [70] designed a MOF-based material MIL-53 (Fe^{II}/Fe^{III}), in which Fe^{II} and Fe^{III} constituted a mixed-valence metal cluster, which mimicked the Fe^{2+} active site and the high-valence metal ions in nitrogen fixation enzymes, respectively. The Fe^{II}/Fe^{III} ratio was crucial for coordinating the catalytic activity and the stability of the backbone structure, and the experimentally obtained optimal Fe^{II}/Fe^{III} ratio was 1.06:1, which gives the highest ammonia yield of 306 µmol·h^{-1}·g^{-1}. The activity enhancement of the MIL-53(Fe^{II}/Fe^{III}) material was attributed to the combined effect between catalytic and non-catalytic functions, i.e., increased ligand-unsaturated active sites, prolonged visible absorption edge (650 nm), and reduced photogenerated carrier complexation rate (Figure 5c).

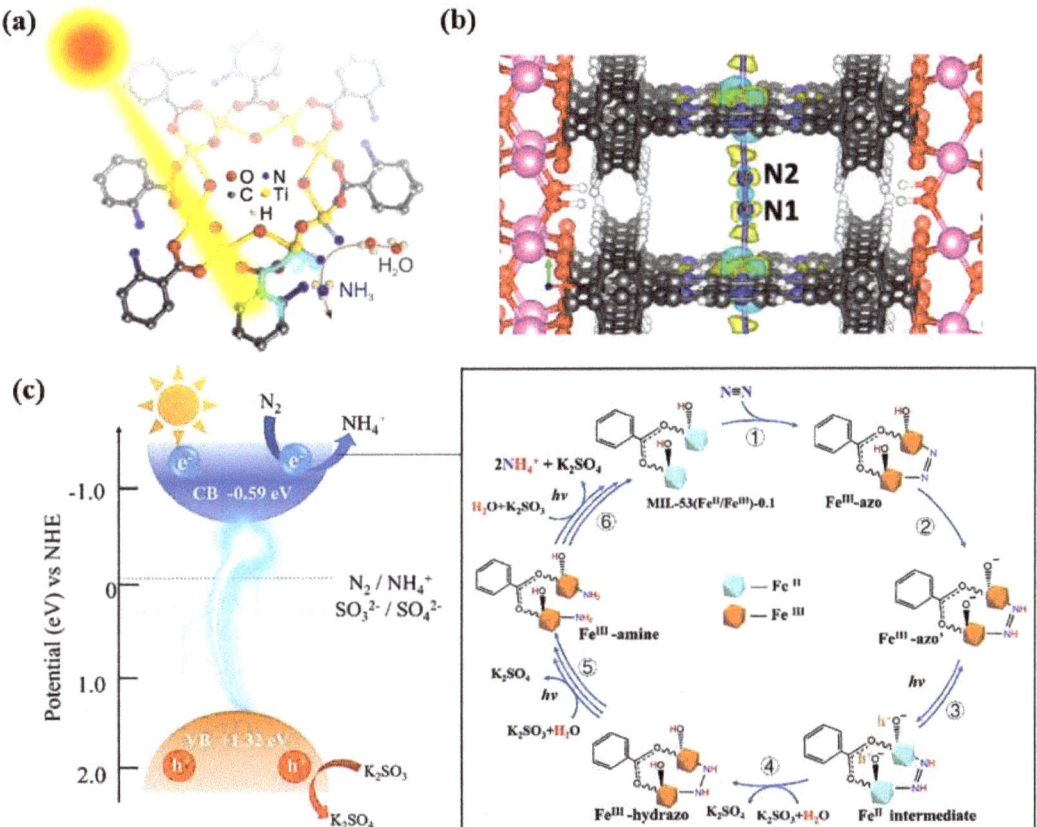

Figure 5. (a) Possible mechanism of NH$_2$-MIL-125(Ti) photocatalytic immobilization of N$_2$ [66]; copyright 2020, Elsevier. (b) Charge difference diagram of N$_2$ adsorbed on AlPMOF (Fe): yellow, positive density difference; cyan, negative density difference [67]. (c) Mechanism of photocatalytic N$_2$ reduction by MIL-53 (Fe^{II}/Fe^{III}) in visible light [70]. Copyright 2020, Elsevier.

Introducing foreign atoms into the main lattice of a semiconductor induces defective states in the electronic and chemical structure, which in turn affects the overall performance of the catalyst [71]. In the photocatalytic nitrogen fixation process, the critical roles of the dopant sites were to act as active centers for N$_2$ adsorption activation and to promote photogenerated charge separation. In addition to the materials mentioned above, Table 2 organizes the recent representative photocatalysts for nitrogen fixation and summarizes the photocatalytic systems by catalyst type, sacrificial agent, light source, and ammonia yield.

Table 2. Recently reported photocatalytic systems for nitrogen fixation.

Type	Photocatalyst	Sacrificial Agent	Light Source	Ammonia Yield/$\mu mol \cdot g_{cat}^{-1} \cdot h^{-1}$	Ref.
Metal oxides	N-TiO$_2$	-	300 W Xe lamp (λ > 400 nm)	80.09	[72]
	Ni-TiO$_2$	-	300 W Xe lamp	46.8	[73]
	Rutile TiO$_2$	Methanol	300 W Xe lamp (λ > 420 nm)	116	[74]
	Fe-TiO$_2$-SiO$_2$	-	300 W Xe lamp	32	[75]
	TiO$_2$-Au-BiOI	-	300 W Xe lamp	534.5	[76]
	Cu$_x$O/CNNTs	Ethanol	300 W Xe lamp (λ > 420 nm)	1380	[77]
	WO$_3$/B-CN	Methanol	300 W Xe lamp (λ > 420 nm)	450.94	[78]
	U-Cu$_2$O-0.05 M-2 h	-	300 W Xe lamp (λ > 400 nm)	4100	[79]
	Cu$_2$O/MoS$_2$/ZnO-cm	-	350 W Xe lamp (λ > 420 nm)	111.94	[80]
Metal sulfides	1T′-MoS$_2$/CNNC	Methanol	300 W Xe lamp (AM 1.5G filter)	9800	[81]
	CdS/CNS	-	350 W Xe lamp (400–800 nm)	327	[82]
	5%NiS-KNbO$_3$	Ethanol	300 W Xe lamp	155.6	[83]
	MoS$_2$/OPC	-	300 W Xe lamp (λ > 400 nm)	37.878	[84]
	Zn$_{0.8}$Cd$_{0.2}$S	Sodium sulfite	300 W Xe lamp (λ > 420 nm)	66.91	[85]
	Fe-MoS$_y$/Cu$_x$S	-	350 W Xe lamp (λ > 420 nm)	8171	[86]
BiOX-based materials	BiVO$_4$	-	300 W Xe lamp (200–800 nm)	103.4	[87]
	P-Bi$_2$WO$_6$	-	300 W Xe lamp	73.6	[88]
	Bi@BiOBr-Bi$_2$MoO$_6$	-	300 W Xe lamp	167.2	[89]
	Fe-BiOCl	-	300 W Xe lamp (200–800 nm)	60	[90]
	Mo-Bi$_5$O$_7$Br-1	-	300 W Xe lamp (λ > 420 nm)	122.9	[91]
	Bi$_5$O$_7$Br	-	300 W Xe lamp (λ > 400 nm)	12700	[92]
Carbon-based materials	S-CNNTs	Ethanol	300 W Xe lamp (λ > 420 nm)	640	[93]
	B-C$_3$N$_5$	Methanol	300 W Xe lamp (200–2500 nm)	421.18	[94]
	Ti$_3$C$_2$/g-C$_3$N$_4$	Methanol	300 W Xe lamp (λ > 420 nm)	601	[95]
	RuPd NPs/C$_3$N$_4$	Ethanol	300 W Xe lamp (λ > 420 nm)	1389.84	[96]
	NYF(15)/NV-CNNTs	Ethanol	300 W Xe lamp (λ > 420 nm)	1720	[97]
MOF-based materials	Zn-MIL-88A	-	300 W Xe lamp	300	[98]
	Au@UiO-66/PTFE membrane	-	300 W Xe lamp (λ > 400 nm)	360	[99]
	MOF@DF-C$_3$N$_4$	Methanol	300 W Xe lamp (λ > 400 nm)	2320	[100]

3. Other Photocatalytic Nitrogen Fixation Materials

In addition to the common photocatalytic materials mentioned above, single-atom catalysts, black phosphorus, layered double hydroxides, molecular sieves, and plasmonic materials have also been shown to have photocatalytic nitrogen fixation activity.

The size of the catalyst directly affects the number of surface low coordination sites, influencing the binding strength to the reactants and determining the catalytic performance to a certain extent. Single-atom metals dispersed on the carrier have the characteristics of uniform catalytic active sites, the low coordination environment of metal atoms, and optimal metal utilization efficiency. Hence, single-atom catalysts have outstanding catalytic activity, stability, and selectivity and have recently attracted wide attention [101]. Liu et al. [102] designed and prepared Ru single-atom modified oxygen-rich vacancy TiO_2 nanosheets, which catalyzed the nitrogen under xenon lamp light reduction to ammonia. The composite photocatalyst containing 1 wt% Ru showed a significantly improved NH_3 generation rate of 56.3 $\mu g \cdot h^{-1} \cdot g_{cat}^{-1}$, two times higher than the performance of the pure TiO_2 nanosheets. DFT calculations showed that the single Ru metal atoms were immobilized on oxygen vacancies, which inhibited the hydrogen precipitation reaction, facilitated the chemical adsorption of N_2, and improved the carrier separation process, resulting in the enhancement of the photocatalytic reduction ability.

Layered double hydroxides (LDHs) belong to two-dimensional nanomaterials, which provide new resources for developing novel catalytic and photocatalytic materials due to their controllable particle size, flexible composition, and easy synthesis. Zhang et al. [103] successfully synthesized ultrathin nanosheets of ZnAl-LDH by a facile co-precipitation method. The 0.5%-ZnAl-LDH nanosheets (Cu doped with 0.5 mol%) with abundant oxygen vacancies and electron-rich ligand unsaturated $Cu^{\delta+}$ exhibited excellent photocatalytic activity and stability under UV–vis irradiation. A catalytic reaction rate of 110 $\mu mol \cdot g^{-1} \cdot h^{-1}$ (4.12 $\mu mol \cdot m^{-2} \cdot h^{-1}$) was achieved at ambient temperature and pressure without any sacrificial agent or co-catalyst addition. Detailed structural analyses and density-functional theory calculations indicate that the oxygen vacancies and $Cu^{\delta+}$ in 0.5%-ZnAl-LDH contribute to the efficient separation and transfer of photogenerated electrons and holes, activating nitrogen molecules and accelerating the multi-electron reduction process.

Plasma catalysis originates from local surface plasmon resonance of metal nanostructures and has been proven to be an effective method for converting light energy into chemical energy. Thanks to the surface plasmon resonance effect of plasma metals and the Schottky barrier formed at the interface with semiconductors, loading plasma metals (Au, Ag, Cu) on semiconductors can effectively expand the light absorption of catalysts to the visible light region and improve the separation efficiency of photogenerated carriers [104]. Xiong's team [105] selected Au nanocrystals to absorb light, and Ru atoms to adsorb N_2 molecules as active sites. They reported a surface plasma that can provide sufficient energy to activate N_2 through a dissociation mechanism in the presence of water and incident light. This mechanism was demonstrated using in situ synchrotron radiation infrared spectroscopy and near-ambient pressure X-ray photoelectron spectroscopy. The photocatalytic nitrogen fixation reaction was carried out using AuRu core-antenna nanostructures with a wide light absorption range and a large number of active sites at room temperature, two atmospheres, and without any sacrificial agents, resulting in an ammonia generation rate of 101.4 $\mu mol \cdot g^{-1} \cdot h^{-1}$. Theoretical simulations have verified that the electric field enhanced by surface plasma, plasma hot electrons, and interface hybridization may play a key role in N≡N dissociation. This work demonstrates the importance of surface plasma in activating inert molecules.

4. Conclusions and Prospects

In the future, the preparation of photocatalysts could be approached by taking into account the following aspects:

In view of the conformational relationship between morphology, structure, and performance, a structural morphology with a large specific surface area could be prepared by changing the reaction conditions, composition, and ratio of precursors to improve the contact probability between the catalysts and the reactants, increase the surface active sites, and promote the adsorption and activation of N_2.

By introducing vacancies, constructing heterojunctions, and element doping, the band gap bandwidth of the semiconductor could be modulated, resulting in enhanced catalyst response to visible light and improved light energy utilization.

To extend the lifetime of photogenerated carriers and to improve the quantum efficiency of photocatalytic reactions, a modification strategy may be utilized to improve the separation and transport efficiency of photogenerated electrons and holes in catalysts. We analyze the mechanisms of photocatalysts and reactions to achieve high energy utilization and excellent catalytic activity in the nitrogen fixation process. We strive to create a reasonable and efficient photocatalytic reaction system tailored for nitrogen fixation.

Author Contributions: C.Z.: writing and revision of the manuscript; Q.S.: conceptualization, methodology, software, investigation, and writing—original draft; L.Y.: funding, acquisition, and supervision. All authors have read and agreed to the published version of the manuscript.

Funding: Financial support for carrying out this work was provided by the Natural Science Foundation of Shandong Province, China (ZR2023QB086).

Institutional Review Board Statement: Not applicable.

Informed Consent Statement: Not applicable.

Data Availability Statement: Not applicable.

Conflicts of Interest: The authors declare no conflict of interest.

Sample Availability: Samples of the compounds are available from the authors.

References

1. Guo, J.P.; Ping, C. Ammonia history in the making. *Nat. Catal.* **2021**, *4*, 734–735. [CrossRef]
2. Wang, M.; Khan, M.A.; Mohsin, I. Can sustainable ammonia synthesis pathways compete with fossil-fuel based Haber-Bosch processes? *Energy Environ. Sci.* **2021**, *14*, 2535–2548. [CrossRef]
3. Smith, C.; Hill, A.K.; Torrente-Murciano, L. Current and future role of Haber-Bosch ammonia in a carbon-free energy landscape. *Energy Environ. Sci.* **2020**, *13*, 331–344. [CrossRef]
4. Wang, L.; Xia, M.K.; Wang, H.; Huang, K.F.; Qian, C.X.; Maravelias, C.T.; Ozin, G.A. Greening ammonia toward the solar ammonia refinery. *Joule* **2018**, *2*, 1055–1074. [CrossRef]
5. Wei, Y.X.; Jiang, W.J.; Liu, Y.; Bai, X.J.; Hao, D.; Ni, B.J. Recent advances in photocatalytic nitrogen fixation and beyond. *Nanoscale* **2022**, *14*, 2990–2997. [CrossRef] [PubMed]
6. Prasidha, K.W.; Widyatama, A. Energy-saving and environmentally-benign integrated ammonia production system. *Energy* **2021**, *235*, 121400.
7. Wang, Y.; Meyer, T.J. A route to renewable energy triggered by the Haber-Bosch process. *Chem* **2019**, *5*, 496–497. [CrossRef]
8. Zuo, C.; Su, Q. Advances in semiconductor-based nanocomposite photo(electro)catalysts for nitrogen reduction to ammonia. *Molecules* **2023**, *28*, 2666. [CrossRef]
9. Rej, S.; Hejazi, S.M.H.; Badura, Z.; Zoppellaro, G.; Kalytchuk, S.; Kment, S.; Fornasiero, P.; Naldoni, A. Light-induced defect formation and Pt single atoms synergistically boost photocatalytic H_2 production in 2D TiO_2-bronze nanosheets. *ACS Sustain. Chem. Eng.* **2023**, *10*, 17286–17296. [CrossRef]
10. Cheng, M.; Xiao, C.; Xie, Y. Photocatalytic nitrogen fixation: The role of defects in photocatalysts. *J. Mater. Chem. A* **2019**, *7*, 19616–19633. [CrossRef]
11. Liang, C.; Niu, H.Y.; Guo, H. Efficient photocatalytic nitrogen fixation to ammonia over bismuth monoxide quantum dots-modified defective ultrathin graphitic carbon nitride. *Chem. Eng. J.* **2021**, *406*, 126868. [CrossRef]
12. Soria, J.; Conesa, J.C.; Augugliaro, V. Dinitrogen photoreduction to ammonia over titanium dioxide powders doped with ferric ions. *J. Phys. Chem. C* **1991**, *22*, 274–282. [CrossRef]
13. Radford, P.P.; Francis, C.G. Photoreduction of nitrogen by metal doped titanium dioxide powders: A novel use for metal vapour techniques. *Chem. Commun.* **1983**, *24*, 1520–1521. [CrossRef]
14. Zhao, W.; Jing, Z.; Xi, Z. Enhanced nitrogen photofixation on Fe-doped TiO_2 with highly exposed (101) facets in the presence of ethanol as scavenger. *Appl. Catal. B-Environ.* **2014**, *144*, 468–477. [CrossRef]
15. Song, G.X.; Gao, R.; Zhao, Z. High-spin state Fe(III) doped TiO_2 for electrocatalytic nitrogen fixation induced by surface F modification. *Appl. Catal. B-Environ.* **2022**, *301*, 120809. [CrossRef]

16. Patil, S.; Basavarajappa, S.B.; Patil, N.G. Recent progress in metal-doped TiO$_2$, non-metal doped/codoped TiO$_2$ and TiO$_2$ nanostructured hybrids for enhanced photocatalysis. *Int. J. Hydrogen Energy* **2020**, *45*, 7764–7778.
17. Li, X.H.; Li, J.X.; Zhai, H.J. Efficient catalytic fixation nitrogen activity under visible light by Molybdenum doped mesoporous TiO$_2$. *Catal. Lett.* **2022**, *152*, 116–123. [CrossRef]
18. Zuo, C.; Tai, X.S.; Su, Q.; Jiang, Z.Y.; Guo, Q.J. S-scheme OV-TiO$_2$@Cu$_7$S$_4$ heterojunction on copper mesh for boosting visible-light nitrogen fixation. *Opt. Mater.* **2023**, *137*, 113560. [CrossRef]
19. Khader, M.M.; Lichtin, N.N.; Vurens, G.H. Photoassisted catalytic dissociation of water and reduction of nitrogen to ammonia on partially reduced ferric oxide. *Langmuir* **1987**, *3*, 303–304. [CrossRef]
20. Licht, S.; Cui, B. Ammonia synthesis by N$_2$ and steam electrolysis in molten hydroxide suspensions of nanoscale Fe$_2$O$_3$. *Science* **2014**, *345*, 637–640. [CrossRef]
21. Sun, S.M.; An, Q.; Wang, W.Z. Efficient photocatalytic reduction of dinitrogen to ammonia on bismuth monoxide quantum dots. *J. Mater. Chem. A* **2017**, *5*, 201–209. [CrossRef]
22. Hou, T.T.; Xiao, Y.; Cui, P.X. Operando Oxygen Vacancies for Enhanced Activity and Stability toward Nitrogen Photofixation. *Adv. Energy Mater.* **2019**, *9*, 1902319. [CrossRef]
23. Li, X.M.; Wang, W.Z.; Jiang, D. Efficient solar-driven nitrogen fixation over Carbon-Tungstic-Acid hybrids. *Chem.-Eur. J.* **2016**, *22*, 13819–13822. [CrossRef] [PubMed]
24. Janet, C.M.; Navaladian, S.; Viswanathan, B. Heterogeneous wet chemical synthesis of superlattice-type hierarchical ZnO architectures for concurrent H$_2$ production and N$_2$ reduction. *J. Phys. Chem. C* **2010**, *114*, 2622–2632. [CrossRef]
25. Nguyen, V.H.; Mousavi, M.; Ghasemi, J.B. High-impressive separation of photoinduced charge carriers on step-scheme ZnO/ZnSnO$_3$/Carbon Dots heterojunction with efficient activity in photocatalytic NH$_3$ production. *J. Taiwan Inst. Chem. E* **2021**, *118*, 140–151. [CrossRef]
26. Zhao, W.R.; Xi, H.P.; Zhang, M. Enhanced quantum yield of nitrogen fixation for hydrogen storage with in situ-formed carbonaceous radicals. *Chem. Commun.* **2015**, *51*, 4785–4788. [CrossRef] [PubMed]
27. Cao, S.; Zhou, N.; Gao, F. All-solid-state Z-scheme 3, 4-dihydroxybenzaldehyde-functionalized Ga$_2$O$_3$/graphitic carbon nitride photocatalyst with aromatic rings as electron mediators for visible-light photocatalytic nitrogen fixation. *Appl. Catal. B-Environ.* **2017**, *218*, 600–610. [CrossRef]
28. Xiao, J.H.; Lv, J.H.; Lu, Q.F. Building Fe$_2$O$_3$/MoO$_3$ nanorod heterojunction enables better tetracycline photocatalysis. *Mater. Lett.* **2022**, *311*, 131580. [CrossRef]
29. Hao, Y.C.; Dong, X.L.; Zhai, S.R. Hydrogenated bismuth molybdate nanoframe for efficient sunlight-driven nitrogen fixation from air. *Chem.-Eur. J.* **2016**, *22*, 18722–18728. [CrossRef]
30. Khan, M.; Bhardwaj, R.C.; Bhardwaj, C. Catalytic fixation of nitrogen by the photocatalytic CdS/Pt/RuO$_2$ particulate system in the presence of aqueous [Ru(Hedta)N$_2$] complex. *Angew. Chem. Int. Ed. Engl.* **1988**, *27*, 923–925. [CrossRef]
31. Ye, L.Q.; Han, C.Q.; Ma, Z.Y. Ni$_2$P loading on Cd$_{0.5}$Zn$_{0.5}$S solid solution for exceptional photocatalytic nitrogen fixation under visible light. *Chem. Eng. J.* **2017**, *307*, 311–318. [CrossRef]
32. Bernardo, I.D.; Blyth, J.; Watson, L. Defects, band bending and ionization rings in MoS$_2$. *J. Phys. Condens. Matter* **2022**, *34*, 174002. [CrossRef]
33. Mao, Y.Y.; Fang, Y.Q.; Yuan, K.D. Effect of vanadium doping on the thermoelectric properties of MoS$_2$. *J. Alloys Compd.* **2022**, *903*, 163921. [CrossRef]
34. Jiang, J.; Chen, Z.Z.; Hu, Y. Flexo-photovoltaic effect in MoS$_2$. *Nat. Nanotechnol.* **2021**, *16*, 894–901. [CrossRef] [PubMed]
35. Sun, S.; Li, X.; Wang, W. Photocatalytic robust solar energy reduction of dinitrogen to ammonia on ultrathin MoS$_2$. *Appl. Catal. B-Environ.* **2017**, *200*, 323–329. [CrossRef]
36. Brown, K.A.; Harris, D.F.; Wilker, M.B. Light-driven dinitrogen reduction catalyzed by a CdS:nitrogenase MoFe protein biohybrid. *Science* **2016**, *352*, 448–450. [CrossRef]
37. Banerjee, A.; Yuhas, B.D.; Margulies, E.A. Photochemical nitrogen conversion to ammonia in ambient conditions with FeMoS-chalcogels. *J. Am. Chem. Soc.* **2015**, *137*, 2030–2034. [CrossRef] [PubMed]
38. Liu, J.; Kelley, M.S.; Wu, W. Nitrogenase-mimic iron-containing chalcogels for photochemical reduction of dinitrogen to ammonia. *Proc. Natl. Acad. Sci. USA* **2016**, *113*, 5530–5535. [CrossRef]
39. Hoffman, B.M.; Dean, D.R.; Seefeldt, L.C. Climbing nitrogenase: Toward a mechanism of enzymatic nitrogen fixation. *Acc. Chem. Res.* **2009**, *42*, 609–619. [CrossRef]
40. John, S.; Anderson, J.R.; Jonas, C.P. Catalytic conversion of nitrogen to ammonia by an iron model complex. *Nature* **2013**, *501*, 84–87.
41. Huang, W.L.; Zhu, Q. DFT calculations on the electronic structures of BiOX (X = F, Cl, Br, I) photocatalysts with and without semicore Bi 5d states. *J. Comput. Chem.* **2009**, *30*, 183–190. [CrossRef] [PubMed]
42. Zhang, M.; Yin, H.F.; Yao, J.C. All-solid-state Z-scheme BiOX(Cl,Br)-Au-CdS heterostructure: Photocatalytic activity and degradation pathway. *Colloids Surf. A* **2020**, *602*, 124778. [CrossRef]
43. Ahern, J.C.; Fairchild, R.; Thomas, J.S. Characterization of BiOX compounds as photocatalysts for the degradation of pharmaceuticals in water. *Appl. Catal. B-Environ.* **2015**, *179*, 229–238. [CrossRef]
44. Li, H.; Shang, J.; Ai, Z.H. Efficient visible light nitrogen fixation with BiOBr nanosheets of oxygen vacancies on the exposed {001} facets. *J. Am. Chem. Soc.* **2015**, *137*, 6393–6399. [CrossRef] [PubMed]

45. Li, H.; Shang, J.; Shi, J.G. Facet-dependent solar ammonia synthesis of BiOCl nanosheets via a proton-assisted electron transfer pathway. *Nanoscale* **2016**, *8*, 1986–1993. [CrossRef] [PubMed]
46. Bai, Y.; Ye, L.Q.; Chen, T. Facet-dependent photocatalytic N_2 fixation of bismuth-rich Bi_5O_7I nanosheets. *ACS Appl. Mater. Interfaces* **2016**, *8*, 27661–27668. [CrossRef] [PubMed]
47. Zeng, L.; Zhe, F.; Wang, Y. Preparation of interstitial carbon doped BiOI for enhanced performance in photocatalytic nitrogen fixation and methyl orange degradation. *J. Colloid Interfaces Sci.* **2019**, *539*, 563–574. [CrossRef]
48. Wang, S.Y.; Hai, X.; Ding, X. Light-switchable oxygen vacancies in ultrafine Bi_5O_7Br nanotubes for boosting solar-driven nitrogen fixation in pure water. *Adv. Mater.* **2017**, *29*, 1701774. [CrossRef]
49. Zhu, D.; Zhang, L.; Ruther, R.E. Photo-illuminated diamond as a solid-state source of solvated electrons in water for nitrogen reduction. *Nat. Mater.* **2013**, *12*, 836–841. [CrossRef]
50. Bandy, J.A.; Zhu, D.; Hamers, R.J. Photocatalytic reduction of nitrogen to ammonia on diamond thin films grown on metallic substrates. *Diam. Relat. Mater.* **2016**, *64*, 34–41. [CrossRef]
51. Tian, Y.; Hu, S.; Sheng, X.L. Non-transition-metal catalytic system for N_2 reduction to NH_3: A density functional theory study of Al-doped Graphene. *J. Phys. Chem. Lett.* **2018**, *9*, 570–576. [CrossRef] [PubMed]
52. Li, X.F.; Li, Q.K.; Cheng, J. Conversion of dinitrogen to ammonia by FeN_3-embedded graphene. *J. Am. Chem. Soc.* **2016**, *138*, 8706–8709. [CrossRef]
53. Chen, S.; Perathoner, S.; Ampelli, C. Room-temperature electrocatalytic synthesis of NH_3 from H_2O and N_2 in a gas-liquid-solid three-phase reactor. *ACS Sustain. Chem. Eng.* **2017**, *5*, 7393–7400. [CrossRef]
54. Liu, Y.; Su, Y.; Quan, X. Facile ammonia synthesis from electrocatalytic N_2 reduction under ambient conditions on N-doped porous Carbon. *ACS Catal.* **2018**, *8*, 1186–1191. [CrossRef]
55. Dong, G.H.; Ho, W.K.; Wang, C.Y. Selective photocatalytic N_2 fixation dependent on g-C_3N_4 induced by nitrogen vacancies. *J. Mater. Chem. A* **2015**, *3*, 23435–23441. [CrossRef]
56. Wu, G.; Gao, Y.; Zheng, B. Template-free method for synthesizing sponge-like graphitic carbon nitride with a large surface area and outstanding nitrogen photofixation ability induced by nitrogen vacancies. *Ceram. Int.* **2016**, *42*, 6985–6992. [CrossRef]
57. Cao, S.H.; Chen, H.; Jiang, F. Nitrogen photofixation by ultrathin amine-functionalized graphitic carbon nitride nanosheets as a gaseous product from thermal polymerization of urea. *Appl. Catal. B-Environ.* **2018**, *224*, 222–229. [CrossRef]
58. Li, X.M.; Sun, X.; Zhang, L. Efficient photocatalytic fixation of N_2 by KOH-treated g-C_3N_4. *J. Mater. Chem. A* **2018**, *6*, 3005–3011. [CrossRef]
59. Hu, S.Z.; Chen, X.; Li, Q. Fe^{3+} doping promoted N_2 photofixation ability of honeycombed graphitic carbon nitride: The experimental and density functional theory simulation analysis. *Appl. Catal. B-Environ.* **2017**, *201*, 58–69. [CrossRef]
60. Yu, L.; Mo, Z.; Zhu, X.; Deng, J.; Xu, F.; Song, Y.; She, Y.; Li, H.; Xu, H. Construction of 2D/2D Z-scheme MnO_{2-x}/g-C_3N_4 photocatalyst for efficient nitrogen fixation to ammonia. *Green Energy Environ.* **2021**, *6*, 538–545. [CrossRef]
61. Nguyen, V.H.; Mousavi, M.; Ghasemi, J.B. In situ preparation of g-C_3N_4 nanosheet/FeOCl: Achievement and promoted photocatalytic nitrogen fixation activity. *J. Colloid Interfaces Sci.* **2021**, *587*, 538–549. [CrossRef] [PubMed]
62. Hu, K.Q.; Huang, Z.W.; Zeng, L.W. Recent advances in MOF-based materials for photocatalytic nitrogen fixation. *Eur. J. Inorg. Chem.* **2022**, e202100748. [CrossRef]
63. Mohamed, A.M.O.; Bicer, Y. The search for efficient and stable metal-organic frameworks for photocatalysis: Atmospheric fixation of nitrogen. *Appl. Surf. Sci.* **2022**, *583*, 152376. [CrossRef]
64. Zhao, C.; Pan, X.; Wang, Z.H. 1 + 1 > 2: A critical review of MOF/bismuth-based semiconductor composites for boosted photocatalysis. *Chem. Eng. J.* **2021**, *417*, 128022. [CrossRef]
65. Shang, S.S.; Xiong, W.; Yang, C. Nano-SH-MOF@Self-Assembling Hollow Spherical g-C_3N_4 Heterojunction for Visible-Light Photocatalytic Nitrogen Fixation. *ChemCatChem* **2023**, *15*, e202201605. [CrossRef]
66. Huang, H.; Wang, X.S.; Philo, D. Toward visible-light-assisted photocatalytic nitrogen fixation: A titanium metal organic framework with functionalized ligands. *Appl. Catal. B-Environ.* **2020**, *267*, 118686. [CrossRef]
67. Shang, S.S.; Xiong, W.; Yang, C. Atomically dispersed iron metal site in a porphyrin-based metal-organic framework for photocatalytic nitrogen fixation. *ACS Nano* **2021**, *15*, 9670–9678. [CrossRef]
68. Zhang, C.; Xu, Y.; Lv, C. Mimicking π backdonation in Ce-MOFs for solar driven ammonia synthesis. *ACS Appl. Mater. Interfaces* **2019**, *11*, 29917–29923. [CrossRef]
69. Zhang, X.; Li, X.M.; Gao, W.G. Bimetallic $CeZr_5$-UiO-66 as a highly efficient photocatalyst for the nitrogen reduction reaction. *Sustain. Energy Fuels* **2021**, *5*, 4053–4059. [CrossRef]
70. Zhao, Z.F.; Yang, D.; Ren, H.J. Nitrogenase-inspired mixed-valence MIL-53(Fe^{II}/Fe^{III}) for photocatalytic nitrogen fixation. *Chem. Eng. J.* **2020**, *400*, 125929. [CrossRef]
71. Li, C.; Gu, M.Z.; Gao, M.M. N-doping TiO_2 hollow microspheres with abundant oxygen vacancies for highly photocatalytic nitrogen fixation. *J. Colloid Interfaces Sci.* **2020**, *609*, 341–352. [CrossRef] [PubMed]
72. Liu, Q.Y.; Wang, H.D.; Tang, R. Rutile TiO_2 nanoparticles with oxygen vacancy for photocatalytic nitrogen fixation. *ACS Appl. Nano Mater.* **2021**, *4*, 8674–8679. [CrossRef]
73. Wu, S.Q.; Chen, Z.Y.; Yue, W.H. Single-atom high-valent Fe(IV) for promoted photocatalytic nitrogen hydrogenation on porous TiO_2-SiO_2. *ACS Catal.* **2021**, *11*, 4362–4371. [CrossRef]

74. Yu, X.J.; Qiu, H.R.; Wang, Z. Constructing of hybrid structured $TiO_2/Au/BiOI$ nanocomposite for enhanced photocatalytic nitrogen fixation. *Appl. Surf. Sci.* **2021**, *556*, 149785. [CrossRef]
75. Zhong, X.; Zhu, Y.X.; Sun, Q.F. Tunable Z-scheme and Type II heterojunction of Cu_xO nanoparticles on carbon nitride nanotubes for enhanced visible-light ammonia synthesis. *Chem. Eng. J.* **2022**, *442*, 136156. [CrossRef]
76. Zhang, K.; Deng, L.Q.; Huang, M.L. Energy band matching WO_3/B-doped g-C_3N_4 Z-scheme photocatalyst to fix nitrogen effectively. *Colloid Surf. A* **2022**, *633*, 127830. [CrossRef]
77. Li, J.X.; Wang, D.D.; Guan, R.Q. Vacancy-enabled mesoporous TiO_2 modulated by nickel doping with enhanced photocatalytic nitrogen fixation performance. *ACS Sustain. Chem. Eng.* **2020**, *8*, 18258–18265. [CrossRef]
78. Zhang, S.; Zhao, Y.X.; Shi, R. Sub-3 nm ultrafine Cu_2O for visible light-driven nitrogen fixation. *Angew. Chem. Int. Ed.* **2021**, *133*, 2584–2590. [CrossRef]
79. Qian, S.; Wang, W.W.; Zhang, Z.S.; Duan, J.H. Enhanced photocatalytic performance of $Cu_2O/MoS_2/ZnO$ composites on Cu mesh substrate for nitrogen reduction. *Nanotechnology* **2021**, *32*, 285706.
80. Xue, Y.J.; Wang, X.Y.; Liang, Z.Q. The fabrication of graphitic carbon nitride hollow nanocages with semi-metal 1T' phase molybdenum disulfide as co-catalysts for excellent photocatalytic nitrogen fixation. *J. Colloid Interfaces Sci.* **2022**, *608*, 1229–1237. [CrossRef]
81. Zhang, H.Z.; Maimaiti, H.; Zhai, P.S. Preparation and photocatalytic N_2/H_2O to ammonia performance of cadmium sulfide/carbon nanoscrolls. *Appl. Surf. Sci.* **2021**, *542*, 148639. [CrossRef]
82. Zhang, W.Q.; Xing, P.X.; Zhang, J.Y. Facile preparation of novel nickel sulfide modified $KNbO_3$ heterojunction composite and its enhanced performance in photocatalytic nitrogen fixation. *J. Colloid Interfaces Sci.* **2021**, *590*, 548–560. [CrossRef] [PubMed]
83. Liu, B.J.; Qin, J.Z.; Yang, H. MoS_2 nano-flowers stacked by ultrathin sheets coupling with oxygen self-doped porous biochar for efficient photocatalytic N_2 fixation. *ChemCatChem* **2020**, *12*, 5221–5228. [CrossRef]
84. Dong, W.Y.; Liu, Y.T.; Zeng, G.M. Crystal phase engineering $Zn_{0.8}Cd_{0.2}S$ nanocrystals with twin-induced homojunctions for photocatalytic nitrogen fixation under visible light. *J. Photochem. Photobiol. A Chem.* **2020**, *401*, 112766. [CrossRef]
85. Su, Q.; Wang, W.W.; Zhang, Z.S.; Duan, J.H. Sustainable N_2 photofixation promoted by Fe-doped MoS_y/Cu_xS grown on copper mesh. *Opt. Mater.* **2022**, *128*, 112373. [CrossRef]
86. Zhang, G.H.; Meng, Y.; Xie, B. Precise location and regulation of active sites for highly efficient photocatalytic synthesis of ammonia by facet-dependent $BiVO_4$ single crystals. *Appl. Catal. B-Environ.* **2021**, *296*, 120379. [CrossRef]
87. Liu, L.; Liu, J.Q.; Sun, K.L. Novel phosphorus-doped Bi_2WO_6 monolayer with oxygen vacancies for superior photocatalytic water detoxication and nitrogen fixation performance. *Chem. Eng. J.* **2021**, *411*, 128629. [CrossRef]
88. Lan, M.; Zheng, N.; Dong, X.L. Facile construction of a hierarchical Bi@BiOBr-Bi_2MoO_6 ternary heterojunction with abundant oxygen vacancies for excellent photocatalytic nitrogen fixation. *Sustain. Energy Fuels* **2021**, *5*, 2927–2933. [CrossRef]
89. Shen, Z.F.; Li, F.F.; Lu, J.R. Enhanced N_2 photofixation activity of flower-like BiOCl by in situ Fe(III) doped as an activation center. *J. Colloid Interfaces Sci.* **2021**, *584*, 174–181. [CrossRef]
90. Chen, X.; Qi, M.Y.; Li, Y.H. Enhanced ambient ammonia photosynthesis by Mo-doped Bi_5O_7Br nanosheets with light-switchable oxygen vacancies. *Chin. J. Catal.* **2021**, *42*, 2020–2026. [CrossRef]
91. Li, P.S.; Zhou, Z.; Wang, Q. Visible Light-driven nitrogen fixation catalyzed by Bi_5O_7Br nanostructures: Enhanced performance by oxygen vacancies. *J. Am. Chem. Soc.* **2020**, *142*, 12430–12439. [CrossRef] [PubMed]
92. Zhu, Y.X.; Zhong, X.; Jia, X.T. Geometry-tunable sulfur-doped carbon nitride nanotubes with high crystallinity for visible light nitrogen fixation. *Chem. Eng. J.* **2022**, *431*, 133412. [CrossRef]
93. Li, K.; Cai, W.; Zhang, Z.C. Modified g-C_3N_5 for photocatalytic nitrogen fixation to ammonia: Key role of Boron in nitrogen activation. *Chem. Eng. J.* **2022**, *435*, 135017. [CrossRef]
94. Liu, W.Z.; Sun, M.X.; Ding, Z.Y. Ti_3C_2 MXene embellished g-C_3N_4 nanosheets for improving photocatalytic redox capacity. *J. Alloys Compd.* **2021**, *877*, 160223. [CrossRef]
95. de Sá, I.F.; Carvalho, P.H.; Centurion, H.A.; Gonçalves, R.V.; Scholten, J.D. Sustainable Nitrogen Photofixation Promoted by Carbon Nitride Supported Bimetallic RuPd Nanoparticles under Mild Conditions. *ACS Sustain. Chem. Eng.* **2021**, *9*, 8721–8730. [CrossRef]
96. Zhu, Y.X.; Zheng, X.L.; Zhang, W.W. Near-infrared-triggered nitrogen fixation over upconversion nanoparticles assembled carbon nitride nanotubes with nitrogen vacancies. *ACS Appl. Mater. Interfaces* **2021**, *13*, 32937–32947. [CrossRef]
97. Ojha, N.; Kumar, S. Tri-phase photocatalysis for CO_2 reduction and N_2 fixation with efficient electron transfer on a hydrophilic surface of transition-metal-doped MIL-88A (Fe). *Appl. Catal. B-Environ.* **2021**, *292*, 120166. [CrossRef]
98. Chen, L.W.; Hao, Y.C.; Guo, Y. Metal-organic framework membranes encapsulating gold nanoparticles for direct plasmonic photocatalytic nitrogen fixation. *J. Am. Chem. Soc.* **2021**, *143*, 5727–5736. [CrossRef]
99. Ding, Z.; Wang, S.; Xue, C. Nano-MOF@defected film C_3N_4 Z-scheme composite for visible-light photocatalytic nitrogen fixation. *RSC Adv.* **2020**, *10*, 26246. [CrossRef]
100. Niu, X.Y.; Zhu, Q.; Jiang, S.L. Photoexcited electron dynamics of nitrogen fixation catalyzed by Ruthenium single-atom catalysts. *J Phys. Chem. Lett.* **2020**, *11*, 9579–9586. [CrossRef]
101. Qiu, P.X.; Xu, C.M.; Zhou, N. Metal-free black phosphorus nanosheets-decorated graphitic carbon nitride nanosheets with C−P bonds for excellent photocatalytic nitrogen fixation. *Appl. Catal. B-Environ.* **2018**, *221*, 27–35. [CrossRef]

102. Liu, S.Z.; Wang, Y.J.; Wang, S.B. Photocatalytic fixation of nitrogen to ammonia by single Ru atom decorated TiO$_2$ nanosheets. *ACS Sustain. Chem. Eng.* **2019**, *7*, 6813–6820. [CrossRef]
103. Zhang, S.; Zhao, Y.X.; Shi, R. Efficient photocatalytic nitrogen fixation over Cu$^{\delta+}$-modified defective ZnAl-layered double hydroxide nanosheets. *Adv. Energy Mater.* **2020**, *10*, 1901973. [CrossRef]
104. Jia, H.L.; Yang, Y.Y.; Dou, Y.R.; Li, F.; Zhao, M.X.; Zhang, C.Y. (Plasmonic gold core)@(ultrathin ruthenium shell) nanostructures as antenna-reactor photocatalysts toward nitrogen photofixation. *Chem. Commun.* **2022**, *58*, 1013–1016. [CrossRef] [PubMed]
105. Hu, C.Y.; Chen, X.; Jin, J.B.; Han, Y.; Chen, S.M.; Ju, H.X.; Cai, J.; Qiu, Y.R.; Gao, C.; Wang, C.M.; et al. Surface plasmon enabling nitrogen fixation in pure water through a dissociative mechanism under mild conditions. *J. Am. Chem. Soc.* **2019**, *141*, 7807–7814. [CrossRef]

Disclaimer/Publisher's Note: The statements, opinions and data contained in all publications are solely those of the individual author(s) and contributor(s) and not of MDPI and/or the editor(s). MDPI and/or the editor(s) disclaim responsibility for any injury to people or property resulting from any ideas, methods, instructions or products referred to in the content.

Review

Recent Advances of Constructing Metal/Semiconductor Catalysts Designing for Photocatalytic CO_2 Hydrogenation

Zhimin Yuan [1], Xianglin Zhu [2,*] and Zaiyong Jiang [1,*]

[1] School of Chemistry & Chemical Engineering and Environmental Engineering, Weifang University, Weifang 261061, China

[2] Institute for Energy Research, School of Chemistry and Chemical Engineering, Jiangsu University, Zhenjiang 212013, China

* Correspondence: zhuxl@ujs.edu.cn (X.Z.); zaiyongjiang@wfu.edu.cn (Z.J.)

Abstract: With the development of the world economy and the rapid advancement of global industrialization, the demand for energy continues to grow. The significant consumption of fossil fuels, such as oil, coal, and natural gas, has led to excessive carbon dioxide emissions, causing global ecological problems. CO_2 hydrogenation technology can convert CO_2 into high-value chemicals and is considered one of the potential ways to solve the problem of CO_2 emissions. Metal/semiconductor catalysts have shown good activity in carbon dioxide hydrogenation reactions and have attracted widespread attention. Therefore, we summarize the recent research on metal/semiconductor catalysts for photocatalytic CO_2 hydrogenation from the design of catalysts to the structure of active sites and mechanistic investigations, and the internal mechanism of the enhanced activity is elaborated to give guidance for the design of highly active catalysts. Finally, based on a good understanding of the above issues, this review looks forward to the development of future CO_2 hydrogenation catalysts.

Keywords: photocatalytic CO_2 reduction; metal/semiconductor; surface plasmon resonance; photocatalysts

1. Introduction

The world's consistent and large-scale dependence on fossil fuels during the past hundred years has had serious repercussions, including energy crises and the deterioration of the global climate, which was caused primarily by increasing atmospheric concentrations of CO_2 [1–9]. Though it has historically been discharged directly into the atmosphere as waste, increasing global interest in transitioning away from the use of fossil fuels now forces us to change our previous conceptions by employing carbon dioxide as a recyclable feedstock for the production of renewable synthetic fuels [10–17]. In recent decades, several methods have developed for carbon dioxide reduction, for example, electrocatalytic [18–21], photocatalytic [22–28], photoelectrochemical (PEC) [29–32], thermocatalytic [33–37], and photothermal [38–40] reduction, etc. These mentioned methods can be driven by clean electric energy or solar energy; therefore, they are considered the most promising strategies. In the case of electrocatalytic reduction [41–44], the two key issues facing electrocatalytic CO_2 reduction are obtaining an inexpensive and a renewable source of electrical energy and the development of highly active CO_2 reduction electrocatalysts. As a result, photovoltaic cells are typically used as the electricity source (thereby allowing for more direct energy conversion from sunlight), and carefully designed electrocatalysts form the cathodes on which CO_2 reduction takes place. Though much research has been conducted to address these issues, high overpotentials, poor selectivity, low electric current densities, catalyst instability, and relatively high costs of renewable electric energy continue to limit the large-scale application of electrocatalytic CO_2 reduction [45]. Photocatalytic CO_2 reduction offers another appealing strategy for reducing CO_2 into fuels by also directly absorbing solar energy as a source of energy while employing water as an electron donor [46–49]. Due to the limitations of low separation efficiency for photogenerated carriers and slow reaction kinetics in both CO_2 reduction and H_2O oxidation, the highest product formation rates reach only micromoles per gram of catalyst per hour (μmol g_{cat}^{-1} h^{-1}) [50]. To overcome those limitations, many studies have been trying to embed the donor–acceptor scheme into the primary photosensitizing module [51–56]. By comparison, thermocatalytic CO_2 reduction has proven an efficient approach for the hydrogenation of CO_2 to form methanol (CH_3OH), methane (CH_4), carbon monoxide (CO), and more at high rates using heterogeneous catalysts [57–62]. However, these processes require higher reaction temperatures (e.g., 200 to 250 °C) and high pressures (5 to 10 MPa) than the previously mentioned methods. Such high temperatures can be disadvantageous, though, when synthesizing products, such as methanol, whose high reaction exothermocity results in diminishing maximum theoretical yields at higher temperatures [63,64]. A related technology, solar thermal catalysis, has also emerged as a promising approach for the synthesis of renewable fuels from CO_2, combining the efficacy of thermocatalysis with a clean source of energy (focused sunlight).

Compared to the electrocatalytic, photocatalytic, and pure thermocatalytic approaches introduced previously, solar thermal reduction methods utilize energy from a large range of the solar spectrum (making use of materials exhibiting intense, broadband light absorption), while also offering high product selectivity relative to photochemical and electrochemical paths [65]. In light of these advantages, solar thermal catalysis has demonstrated its potential for use in industrially relevant chemical reactions, such as reducing nitrogen to ammonia (Haber process) [66], methane reforming [67], and preparing synthesis gas [68,69]. Though there have been many recent discoveries and developments regarding solar thermal catalytic reactions, these specific examples are highlighted for being significantly endothermic; as such, they benefit greatly from the high temperatures and thermal input generated by solar thermal systems.

2. Application of Metal/Semiconductor Catalysts for Photocatalytic CO_2 Hydrogenation

The solar energy utilization efficiency in photocatalytic processes is determined by both the incident quantum absorption capability of the catalyst and its internal quantum yield. As a result, expanding the range of wavelengths absorbed by a catalyst has been proven to be one of the most effective ways of enhancing catalytic activity, especially for the semiconductor-based materials. Though great attention has been paid to broadening optical absorption ranges, such as introducing dopants, defects, sensitizers, or plasmonic metal, these strategies often result in unintended side effects (weakened redox capacity, decreased internal quantum yield, etc.). Thus, it is necessary to further explore novel catalytic systems and new mechanisms that can maximize both the light absorption efficiency and internal quantum yield [70]. Loading noble metal nanoparticles onto photocatalysts can enhance energy harvesting from visible-light photons, thanks to high-energy ("hot") electrons generated by localized surface plasmon resonance (LSPR) [71,72]. The design of metal/semiconductor catalysts has been one potential strategy for the preparation of efficient photocatalytic CO_2 hydrogenation catalysts.

2.1. Pd-Based Catalysts

To lower the temperature needed in a thermocatalytic CO_2 reduction and increase solar energy utilization, Ni et al., fabricated a palladium-nanoparticle-loaded TiO_2 (PNT) catalyst employed using a photothermochemical cycle (PTC) method [73]. In this system, the uniformly dispersed Pd nanoparticles achieved LSPR absorption in the visible region. As a result, the PNT exhibited enhanced light utilization thanks to its red-shifted absorption range in the visible spectrum (Figure 1a). Absorbed visible IR light could then be converted to thermal energy and incorporated into a chemical reaction, allowing the 1.0 PNT catalyst to realize a stable CO production rate of 11.05 μmol g_{cat}^{-1} h^{-1} (more than eight times the CO production rate of P25 using the PTC) (Figure 1c). The structure of the PNT catalyst also allowed the electron–hole pairs generated from the photoexcitation of TiO_2 to be more efficiently separated (Figure 1b), thereby increasing the number of available charge carriers. Thus, the reduction rate of CO_2 production can be accelerated by an advantage of photothermal synergy. Complementary density functional theory (DFT) calculations indicated that Pd can play a further role in promoting CO_2 adsorption by forming $Pd\text{-}CO_2^-$ and $Pd\text{-}CO_2^-\text{-}V_O$ at defect sites during the thermal catalytic progress (Figure 1d).

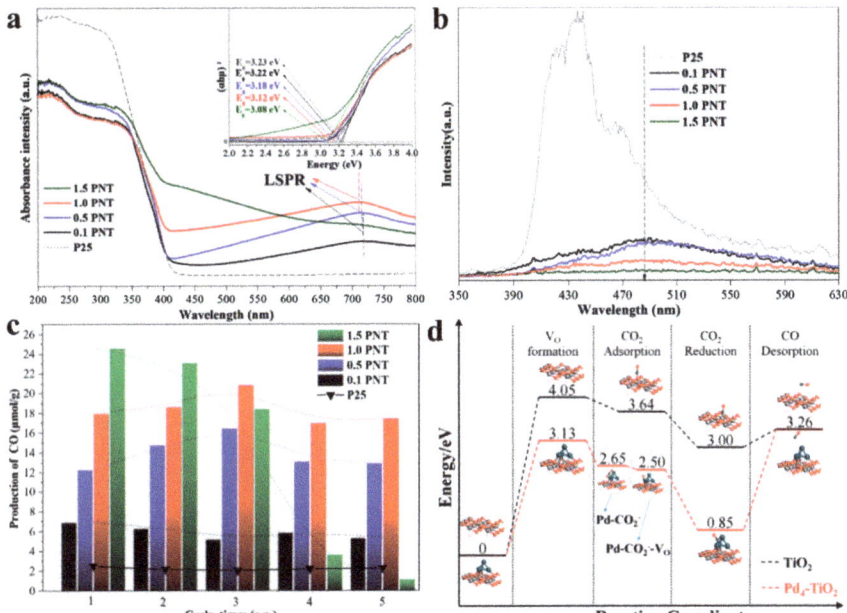

Figure 1. (**a**) Diffuse reflectance UV-vis spectra of original P25 and PNT samples (inset: determination of the optical band gap). (**b**) Photoluminescence spectra of P25 and the PNT samples. (**c**) CO yields obtained from five successive PTC runs. All samples were illuminated for 50 min under a He atmosphere and then heated at 773 K for 50 min under a CO_2 atmosphere (The dashed line represents the changing trend of CO yield for each catalyst.). (**d**) DFT-calculated reaction pathways proceeding on P25 and PNT catalysts [73].

Another promising series of photothermal catalysts was developed by Ozin et al., using Pd to provide deeper insight into the synergistic mechanisms responsible for the highly selective reduction in CO_2. In the first such work, nanostructured $Pd@Nb_2O_5$ nanorod catalysts were prepared via an impregnation method and tested in the gas-phase hydrogenation of CO_2 to CO [74]. The use of ethanol as a reducing agent was instrumental in forming uniformly small Pd nanoparticles on the Nb_2O_5 nanorods, as it allowed the reduction reaction rate of the palladium precursor to be controlled, thereby suppressing particle formation. Experiments proved that the Pd nanoparticles were well dispersed on the surface of the Nb_2O_5 nanorods and had diameters ranging from 2 to 10 nm. Due to their size, these Pd nanocrystals were able to absorb shorter wavelengths of light (visible and near-infrared (NIR) light) by exciting inter-band and intra-band electronic excitations, helping to initiate CO_2 reduction. This increased amount of thermal energy was beneficial, as it was also proven that the reaction rate was significantly enhanced by the generation of Nb^{4+} and oxygen vacancies in situ (on reduction by H_2). The Nb_2O_5 nanorods also featured a large surface area (as measured via BET gas adsorption), which provided a large number of CO_2 adsorption sites for the reaction. Wavelength- and intensity-dependent catalytic activity measurements performed on the nanostructured $Pd@Nb_2O_5$ catalyst (Figure 2) revealed that the reverse water–gas shift (RWGS) reaction was activated photothermally, depending primarily on the conversion on light to thermal energy rather than the formation of electron–hole pairs. By virtue of this combination of features, the RWGS reaction can be catalyzed by the Pd@Nb$_2$O$_5$ catalyst under visible and NIR light even without external heating, achieving an impressive reaction rate of 1.8 mmol g$_{cat}^{-1}$ h^{-1}.

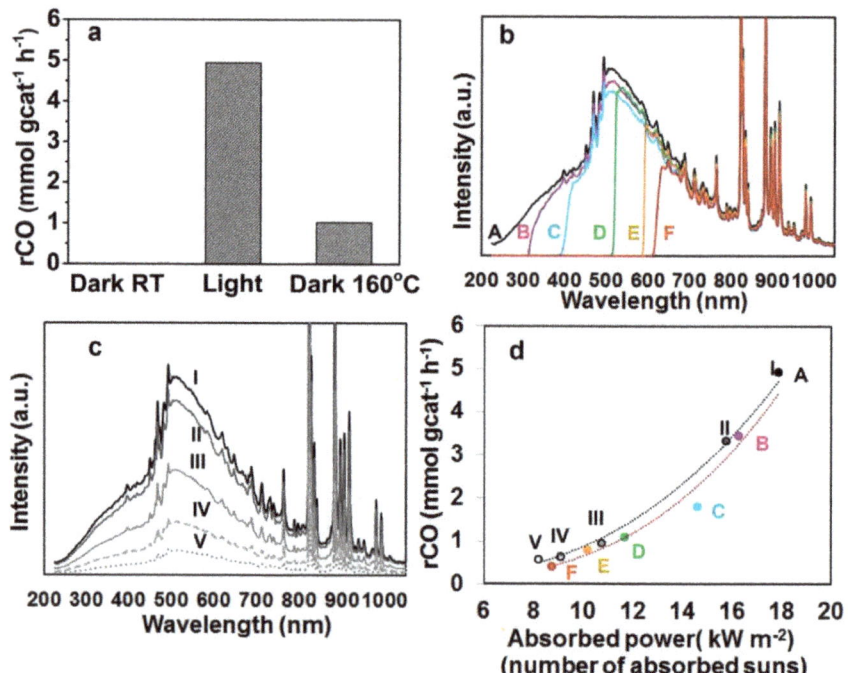

Figure 2. (**a**) Photothermal CO production rates over Pd@Nb$_2$O$_5$. (**b**) Spectral irradiance incident on the Pd@Nb$_2$O$_5$ catalyst with different cut-off filters (A through F). (**c**) Spectral irradiance incident on the Pd@Nb$_2$O$_5$ catalyst for batch reactions I through V. (**d**) RWGS reaction rates plotted as a function of absorbed power for the series of batch reactions A through F (red line) and *I* through *V* (black line) [74].

Silicon-based support materials have also been used in conjunction with ultrasmall (19-atom) Pd nanoparticles, which were deposited onto two-dimensional silicon–hydride nanosheets (Pd@SiNS) and used to catalyze the RWGS reaction. Due to the innovative synthesis method, Pd precursors were reduced in situ on the surface of the Si nanosheets, resulting in highly dispersed and well-immobilized Pd nanoparticles while also retaining the large surface area (197 m^2 g^{-1}) of the two-dimensional Si nanosheets. Several characterization techniques were also employed, including Fourier transform infrared spectroscopy, in situ isotopic labeling, and theoretical calculations to understanding the internal catalytic mechanism. [40] The Pd@SiNS sample showed a ^{13}CO production rate of 10 µmol g$_{cat}^{-1}$ h^{-1}, which was much better than that of the pristine SiNS sample (Figure 3a), and retained its good stability under repeated cycling (relative to pristine SiNSs). A crucial supporting experiment showed that Pd nanoparticles on fully oxidized Si nanosheets (Pd@oxSiNS) had no activity in CO$_2$ reduction (Figure 3b), demonstrating the limited activity of Pd nanoparticles alone and proving the synergetic effect between Pd and the Si nanosheets. By studying these experimental results, a clear and fundamental understanding was reached, in that the H-transfer process from Pd to the oxidized SiNS surface represented two mechanisms. First, one H atom adsorbed onto a Pd nanoparticle interacted with a surface Si-O-Si site of the SiNSs. Resulting in the formation of a surface Si–OH site. Second, another active H from the Pd nanoparticle forms a bond with the generated Si-OH, leading to the desorption of a molecule of water and the formation of a highly reactive surface Si radical site capable of reacting with CO_2 and enabling a catalytic reaction cycle. The strain induced in the SiNS surface by the Si-O-Si bonds is believed to be responsible for the enhanced reactivity of the oxidized SiNS surface toward catalytic hydrogenation of CO_2 under mild conditions. This work also highlights the potential role of Si surface chemistry in designing and developing new heterogeneous CO_2 catalysts in the future.

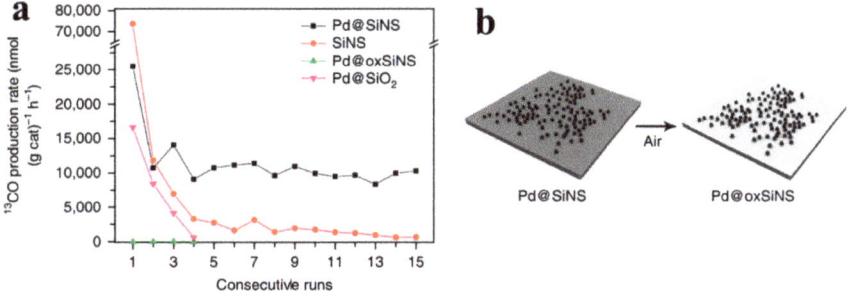

Figure 3. (a) ^{13}CO production rate of four samples: Pd@SiNS, pristine SiNS, Pd@oxSiNS, and Pd@SiO_2. (b) Preparation of the cAontrol sample Pd@oxSiNS via oxidation in air [40].

Similar synergistic mechanisms between metal co-catalysts and metal oxide supports were also presented by Li et al., in which a catalyst consisting of palladium nanoparticles deposited into tungsten–bronze nanowires (Pd@H_yWO_{3-x}) was used to efficiently catalyze the reduction in CO_2 to CO at a rate of 3.0 mmol g_{cat}^{-1} hr^{-1} with a selectivity exceeding 99% [75]. WO_3 nanowires loaded with Pd nanocrystals provided the synthetic pathway to obtaining Pd@H_yWO_{3-x}. Through various characterization techniques, it was demonstrated that H_yWO_{3-x} was formed via spillover of activated H from the Pd nanoparticles. The existence of Brønsted acid hydroxyls (-OH), W(V) sites, and oxygen vacancies (V_O) in H_yWO_{3-x} all contributed to the CO_2 capture and reduction reactions.

2.2. Ru-Based Catalysts

Ye et al., prepared a series of Group VIII metals (Ru, Rh, Ni, Co, Pd, Pt, Ir, and Fe) loaded onto Al_2O_3 or TiO_2 to form classical metal/metal oxide nanocatalysts and studied their photothermal effects on the catalytic reduction in CO_2 into CH_4 and CO (using H_2) [39]. The loading of Group VIII metals onto Al_2O_3 or TiO_2 resulted in highly active materials that had a dark grey or black color (Figure 4a) due largely to their nanoscale structure. A high product selectivity (over 95%) for either CH_4 or CO could be achieved by adjusting the metal and support materials used in a catalyst (Figure 4b). The outstanding CO_2 conversion ability of the prepared catalyst was attributed to two main features: highly efficient utilization via photothermal processes and the unique ability of the Group VIII metals to activate CO_2 hydrogenation. Compared to pure photocatalytic pathways, photothermal CO_2 conversion is not restricted to utilizing ultraviolet irradiation, as it can readily absorb energy from visible and infrared wavelengths (this range contains 96% of the solar energy) can utilized by photothermal CO_2 conversion. In related research from the same group, Ru-loaded ultrathin, layered double hydroxides (Ru@FL-LDHs) were also prepared, achieving efficient photothermal CO_2 methanation in a flow reactor system [76]. These catalysts were composed of well-dispersed Ru nanoparticles embedded in ultrathin, exfoliated LDH sheets. Under light irradiation, a maximum CH_4 production rate of 277 mmol g_{cat}^{-1} h^{-1} or 230.8 mol h^{-1} m^2 (irradiation area) was attained, which was significantly higher than most reported LDH-based catalysts [77–79]. These experiments proved that ultrathin LDH sheets can provide an abundance of active sites for the adsorption and activation of CO_2 molecules, allowing for the use of a high flow rate of feedstocks without decreasing the percent conversion of CO_2 to products. Meanwhile, the highly dispersed Ru nanoparticles contributed by providing higher local surface temperatures under light irradiation (via photothermal heating) and activating H_2 molecules, thereby promoting the hydrogenation of CO$_2$. In this example, the noble metal particles and supporting LDH matrix activated the CO$_2$ and H$_2$ reagents separately, highlighting the potential benefits of designing catalysts by targeting the activation of individual reactant molecules.

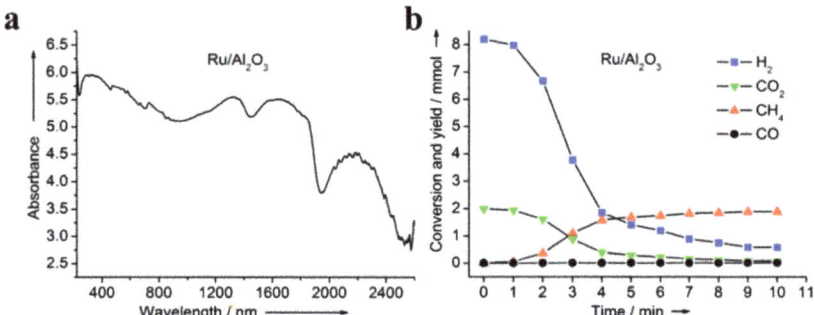

Figure 4. (**a**) Representative UV-Vis–NIR spectrum and (**b**) product yields of a metal/metal oxide photothermal nanocatalyst (Ru/Al$_2$O$_3$) [36].

In a related work by Dai et al., a Ru/TiO$_{(2-x)}$N$_x$ catalyst was prepared by loading Ru onto nitrogen-doped TiO$_2$ support via an impregnation–reduction method [80]. The catalytic CO$_2$ methanation performance of this catalyst was evaluated under both visible light irradiation (435 nm < λ < 465 nm) and under dark conditions in a fixed-bed flow reactor. With the illumination of visible light, the turnover frequency (TOF) of the catalyst increased to 15.0 h^{-1} (compared to the 7.0 h^{-1} under dark condition), with the CH$_4$ selectivity remaining stable at about 80% throughout the duration of the reaction. A combination of chemisorption, surface analysis, and photocurrent measurements revealed that visible light contributed to the catalytic reaction in two ways. First, a greater number of oxygen vacancies were formed due to the stronger visible light response of TiO$_{2-x}$N$_x$, allowing CO$_2$ to be more efficiently adsorbed, activated, and converted into the CO. Second, Ru nanoparticles can accept photogenerated electrons from TiO$_{(2-x)}$N$_x$, increasing the surface electron density of Ru nanoparticles and promoting the adsorption and activation of CO$_2$.

In another study from the same group, Ru/silicon nanowire catalysts (Ru/SiNW) (Figure 5a,b) were prepared by forming 10 nm Ru on silicon nanowires (SiNWs) via sputtering. In this case, the coexistence of Ru and SiNWs resulted in the possibility of activating the Sabatier reaction (CO$_2$ + H$_2$ → CH$_4$ + H$_2$O) both thermochemically and photochemically [81]. Detailed investigations were performed in the presence and absence of light irradiation (Figure 5c), with varying atmospheric pressure, (Figure 5d,e) and using varying wavelengths of light (Figure 5f) to better understand the internal mechanisms responsible for the observed catalytic activity. The results of these investigations revealed that the observed CO$_2$ reduction mechanism could be predominantly photochemical or thermochemical, depending on the wavelengths of irradiating light. When the photon energy was less than the bandgap energy of silicon, the Ru/SiNW catalyst relied upon photothermal heating to activate the Sabatier reaction thermochemically. When the incident photon energy is sufficient to overcome the bandgap of Si, however, electron–hole pairs generated in the Ru/SiNW catalyst assisted the formation of R-H bonds. Eventually, a proportion of these excited charge carriers eventually thermalize and recombine, generating heat and supporting the thermochemical promotion of the Sabatier reaction. In this way, the prepared Ru/SiNW catalysts achieved a CH$_4$ production rate of 0.74 mmol g$_{Ru}^{-1}$ h^{-1} with a simulated solar light intensity of 14.5 suns, a 4:1 H$_2$/CO$_2$ ratio, and an atmospheric pressure of 15 psia. The concepts and the experimental approaches introduced in this work are significantly helpful for understanding the synergistic contributions of photochemical and thermochemical mechanisms to the overall process of photothermal CO$_2$ reduction.

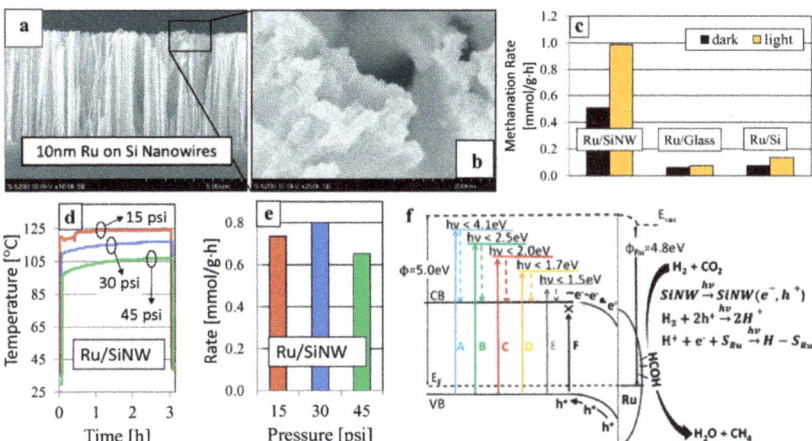

Figure 5. (**a**,**b**) Scanning electron microscopy (SEM) images of 10 nm Ru nanoparticle/silicon nanowires. (**c**) Methane production rates over Ru-based catalysts on SiNW, glass, and polished Si supports (**d**) Temperature profiles recorded for batch reactions performed at a different reaction condition and (**e**) methanation rates of a different sample. (**f**) The proposed energy band diagram at the SiNW-Ru interface and schematic diagram [81].

In yet another study, the photothermal reduction in gaseous CO_2 over Ru/silicon photonic crystal photocatalyst (Figure 6a,b) was studied at ambient temperature [82]. Ru/i-Si-o catalysts showed increasing activity as light intensity increased, achieving peak rates of up to 4.4 mmol g_{cat}^{-1} h^{-1} (Figure 6c). The high photomethanation rates over the Ru/i-Si-o catalyst are attributed to the excellent light-harvesting properties of the silicon photonic crystal, as proven using control experiments (Figure 6d). Complementary simulation results further indicated that the charged Ru surfaces are beneficial for CO_2 destabilization–adsorption progress; additionally, the charged Ru surfaces are crucial for the adsorbing and dissociating of H_2, which can react with CO_2 to accelerate the Sabatier reaction.

2.3. Other Metal-Based Catalysts

Recently, the plasmonic properties of Rh and Au in photocatalytic CO_2 reduction were investigated by Liu et al. [83]. A Rh/Al_2O_3 catalyst was prepared via an impregnation method. Rh nanocube colloids were dropped onto Al_2O_3 nanoparticles, and the formed solid was ground and calcined in air (Figure 7a) and then tested in a fixed-bed reactor (Figure 7b). It was found that the CH_4 and CO production rates for Rh/Al_2O_3 were similar at 350 °C without light irradiation. When ultraviolet light (3 W cm^{-2}) was introduced, a seven-fold increase in the CH_4 production rate was detected, while the CO production rate changed only very slightly (Figure 7c). In contrast, plasmonic Au nanoparticles could catalyze only CO production, regardless of the presence of irradiation (Figure 7d). It is also worth noting that the CH_4 selectivity of unheated Rh nanoparticles was over ~86% or ~98% compared to the dark condition or thermo condition, respectively. And, the reaction rate was double that of the thermocatalytic reaction rate at 350 °C (Figure 7d,e). These experimental results together indicated the selective production of CH_4 via a photothermal catalytic process. Further kinetic studies, together with DFT calculations, revealed that photo-excited hot electrons generated through the LSPR effect of Rh nanocubes would selectively be transferred into the antibonding orbital of the CHO intermediate, which can lower the activation energy of CH_4 production by ~35% relative to the thermal activation energies (Figure 7f). This study well demonstrates the idea of tuning product selectivity in CO_2 reduction by using light to excite plasmonic electrons and activate key reaction intermediates.

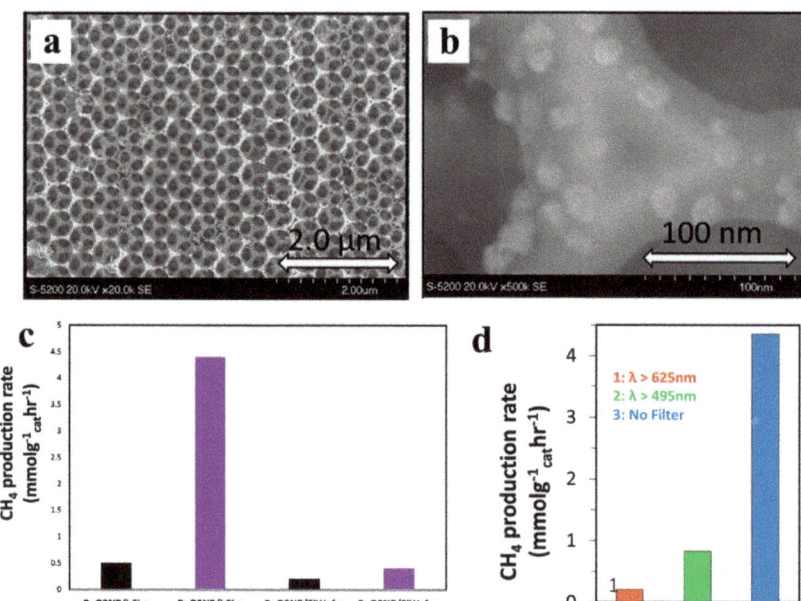

Figure 6. (**a**,**b**) SEM secondary electron images of ncRuO$_2$/i-Si-O sample. (**c**) ^{13}CH$_4$ production rate from ncRuO$_2$/i-Si-O and ncRuO$_2$/Si wafer hybrid samples. (**d**) Photomethanation rates over the ncRuO$_2$/i-Si-o catalyst tested under different illumination conditions [82].

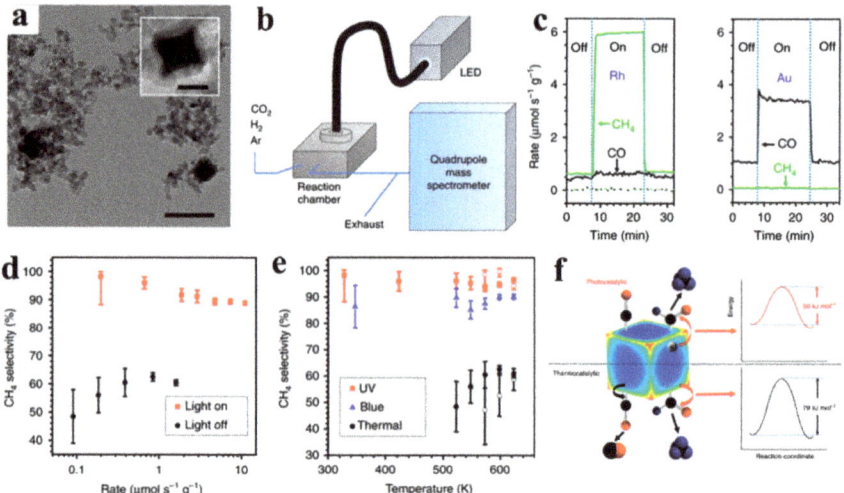

Figure 7. (**a**) Transmission electron microscopy (TEM) images of the Rh/Al$_2$O$_3$ photocatalyst (Scale bar, 100 nm (inset: 25 nm)). (**b**) Schematic of the photocatalytic reaction system. (**c**) Rates of CH$_4$ (green) and CO (black) production at 350 °C on Rh/Al$_2$O$_3$ and Al$_2$O$_3$ under conditions of dark and ultraviolet illumination at 3 W cm^{-2}. (**d**) Selectivity toward CH$_4$ as a function of overall reaction rates in the dark and under ultraviolet light. (**e**) Selectivity toward CH$_4$ of the thermocatalytic and photocatalytic reactions. (**f**) Reaction mechanism on a rhodium nanocube (The black, red, and blue spheres represent carbon atom, oxygen atom, and hydrogen atom.) [83].

In order to achieve an efficient dry reforming reaction ($CO_2 + CH_4 \rightarrow 2CO + 2H_2$), Ye et al., used plasmonic nanoparticles to promote the activity of their noble metal-based catalyst (Figure 8) [9]. In this case, a Rh-Au/SBA-15 bimetallic catalyst was synthesized via a subsequent impregnation pathway, with the prepared catalyst showing a strong LSPR corresponding to Au nanoparticles (Figure 8a) [11]. In a flow reactor system with visible light illumination of 420 to 800 nm and a temperature of 500 °C, the activity catalyst was enhanced 1.7 times compared to that observed in the dark. Additionally, the catalytic activity of Rh-Au/SBA-15 was much higher than that of either Rh/SBA-15 or Au/SBA-15 (Figure 8b). UV-Vis spectra and electromagnetic field simulation results confirmed the corresponding relationship between the enhanced dry reforming activity and the LSPR band of Au (Figure 8c,d). This demonstrated the photothermal activity of the dry reforming reaction over the Rh-Au/SBA-15 catalyst and the role of energetic "hot" electrons excited by LSPR in Au, contributing both thermal energy and polarization/activation of CO_2 and CH_4 to the catalytic reaction.

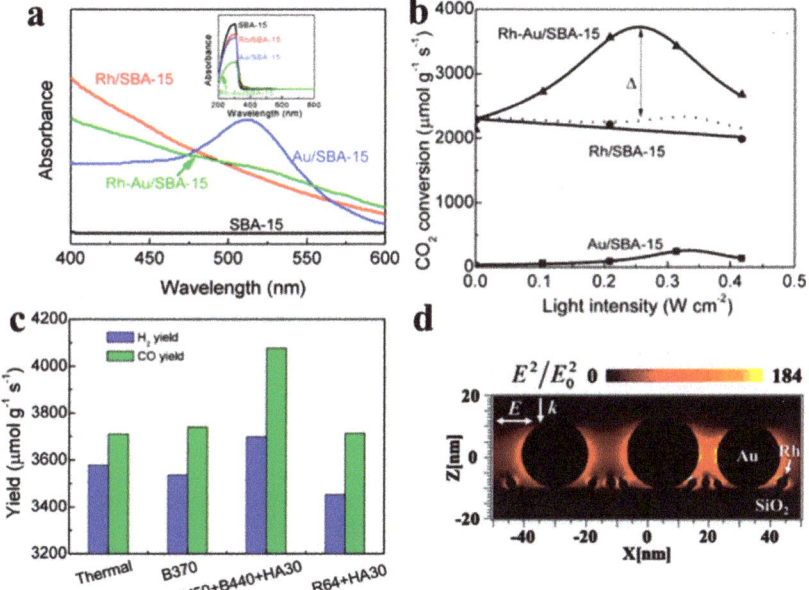

Figure 8. (**a**) UV-Vis spectra of the Rh/SBA-15, Au/SBA-15, and Rh-Au/SBA-15 catalysts. (**b**) Effects of visible light intensity on the performance of catalysts in the dry reforming reaction. (**c**) H_2 and CO yields without and with light irradiation of different wavelength ranges. (**d**) Cross-sectional views of the electromagnetic field distribution and enhancement simulated for Rh-Au/SBA-15 through a finite-difference time-domain (FDTD) method [11].

In another work from the same group, TaN was used as a support for loading Pt nanoparticles, chosen due to its intense light absorption over a range of wavelengths (Figure 9a) [84]. This Pt/TaN catalyst was prepared via a simple impregnation method, resulting in an average Pt nanoparticle diameter of approximately 2.7 nm (Figure 9b). The Pt/TaN catalyst was also tested in the dry reforming reaction under visible light irradiation at 500 °C, which revealed that the CO_2 conversion rate of the catalyst was 2.7 times that observed in the dark. In a control experiment employing Pt/Ta_2O_5 as a catalyst, the catalytic activity remained nearly constant even under visible light illumination (Figure 9c). Using DFT calculation results, the polarity of TaN was revealed to be a critical parameter for the efficient separation of electron–hole pairs generated in the photothermal excitation. In conclusion, this study emphasized the importance of a support material's optical absorption properties in photothermal reactions (Figure 9d).

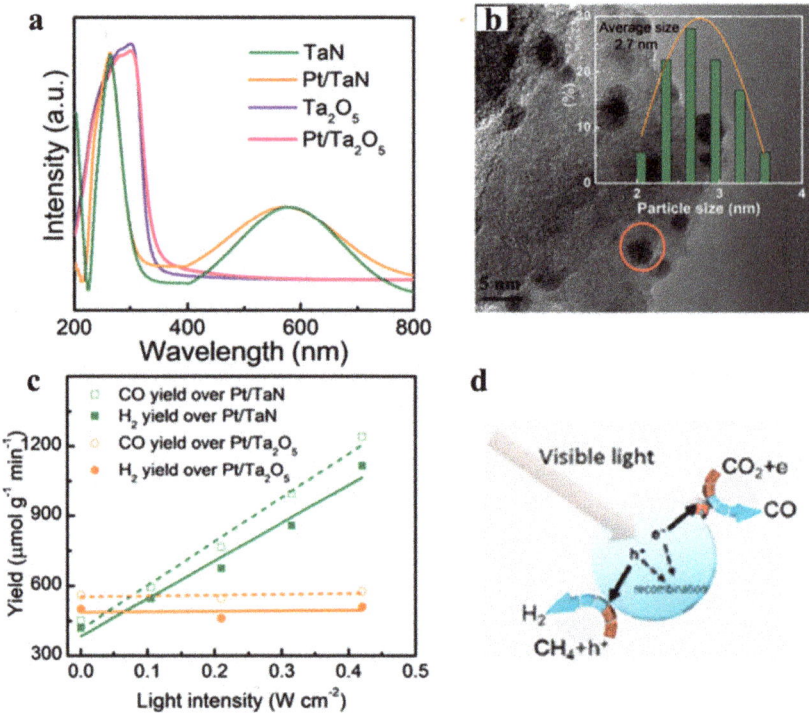

Figure 9. (**a**) UV-Vis spectra of TaN, Ta$_2$O$_5$, and the Pt-based catalysts. (**b**) TEM images of Pt/TaN. (**c**) Catalytic performance of Pt/TaN and Pt/Ta$_2$O$_5$ in the dry reforming reaction. (**d**) Mechanistic illustration of the visible-light-assisted dry-reforming reaction over Pt/TaN catalysts [84].

The aforementioned plasmonic properties of Pd, Rh, and Au demonstrate the promotion of photocatalytic activity by extending the light absorption range, suppressing the charge-carrier recombination, and forming highly active surface reaction sites. However, the relatively high cost and low abundance of noble metals place a hard limit on their practical application on a global scale. With this in mind, Ozin et al., have recently developed plasmonic cobalt superstructures capable of achieving nearly 100% sunlight absorption while acting as photocatalysts for efficient CO$_2$ reduction. [70] This cobalt plasmonic superstructure was composed of a nanoporous, needle-like structure containing Co metal cores and an enclosing silica shell, Co-PS@SiO$_2$ (Figure 10a). For comparison, Co@SiO$_2$ and Co/FTO samples were also prepared via traditional impregnation methods. The diffuse reflectance spectra of different catalysts were characterized as shown in Figure 9b. It was demonstrated that the silica shell can prevent the nanoneedle structure from collapsing and confine the growth of Co nanocrystal. As a result, more active sites will be preserved, and the photon path length can be simultaneously increased through multiple reflection. Through detailed experimental comparisons, this study concluded that the strong light absorption of Co metal combined with the antireflection of the nanoarray structure contributes to the broadband optical absorption across the entire solar spectrum. (Figure 10b). The photocatalytic activity was evaluated in a batch reactor system operating at atmospheric pressure with a 1:1 feed ratio of H$_2$/CO$_2$. The catalytic testing results of the three aforementioned catalysts are presented in Figure 10c,d. The CO$_2$ conversion rate of Co-PS@SiO$_2$ was about 0.6 mol g$_{Co}^{-1}$ h^{-1} (normalized to the mass of Co), which is 6 times that of Co@SiO$_2$ and 277 times that of Co/FTO (Figure 9a). When normalized by surface area, these respective activity increases were 20 and 151 times greater (Figure 10b). Moreover, the CO selectivity of Co-PS@SiO$_2$ was higher than that of Co@SiO$_2$, which is also consistent with the higher local catalyst temperature caused by the better light absorption ability. The ability of Co-PS@SiO$_2$ to fully harvest solar energy demonstrates the applicability of base metal plasmonic superstructures in other solar energy harvesting systems.

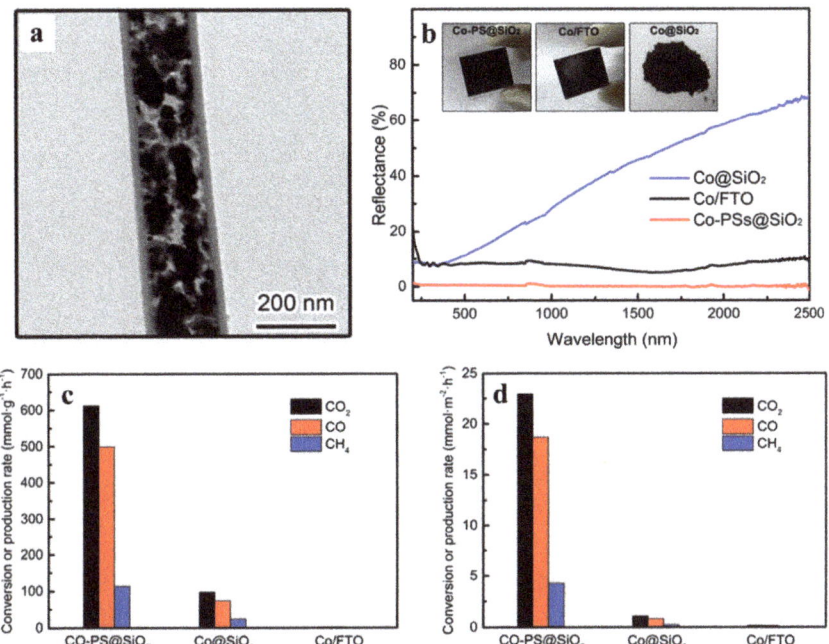

Figure 10. (**a**) Cross-sectional TEM images of Co-PS@SiO$_2$. (**b**) Diffuse reflectance spectra of each cobalt catalyst material. Insets show the corresponding photographs of these samples. (**c**,**d**) Conversion rates of reagents and production rates of products from the catalyst normalized by (**c**) the mass of Co and (**d**) the surface area of Co [70].

3. Summary and Outlook

In this review, we have focused on recent developments in solar-energy-driven CO$_2$ hydrogenation reactions, from the design of catalysts to the structure of active sites and mechanistic investigations. To date, several strategies have been developed to improve the catalytic efficiency and overcome the challenges facing CO$_2$ reduction. For the purpose of utilizing light energy, localized surface plasmon resonances of metal particles (Pd, Rh, Ni, Co, etc.) and vacancies have been proven as two simple and efficient approaches to widening the light absorption range of light-driven catalyst materials. The presence of metal particles not only enhances the optical absorption capability but also provides active sites for the activation of H$_2$ and CO$_2$. As a surface catalytic reaction, the presence of anion vacancies can both improve the adsorption of CO$_2$ molecules and lower the activation energy associated with their reduction. Specialized nanostructures have also been constructed to increase the utilization of light, including photonic crystals that can enhance light trapping. Indium-based oxides are found to have excellent performance in CO$_2$ hydrogenation because of their flexibility in terms of morphologies, phases, and surface-active sites. Though good progress has been made, several challenges must still be overcome before practical applications of these catalysts becomes possible.

First, the catalytic activity of many reported catalysts is still limited; therefore, improving nanoscale architectural design, bandgap engineering, and broad spectral adsorption appears to be particularly important. Many experimental and theoretical studies have been undergoing to study the relationship between spectral adsorption properties and catalytic performance; however, the lack of enough in-situation or operando activities make it a challenge to discover the internal mechanism.

Second, the cost of many of these catalysts is prohibitively high. Most of the developed catalysts contain precious metals; however, increasing atomic efficiency through alloying or the development of single atom catalysts may be offering some improvement. Moreover, it is also important to design some non-noble metals and noble metals (bimetal)/semiconductor catalysts in the process of photocatalytic CO_2 reduction, such as Cu, Bi, et al.

Finally, the industrial-scale preparation of these catalysts and the developments of specialized equipment are still relatively scarcely reported. In a word, with the continuous development of broadband, highly absorbing materials, and increased understanding of photochemical and thermal properties in nanostructured catalysts, the future of photothermal catalysis offers many interesting opportunities for CO_2 capture and conversion technologies. The combining of photocatalytic and thermocatalytic approaches will doubtlessly provide new development opportunities for efficient CO_2 hydrogenation.

Author Contributions: Z.Y. Design, Writing and editing original draft. X.Z.: Design, Writing and editing original draft. Z.J.: Design, Writing and editing original draft. All authors have read and agreed to the published version of the manuscript.

Funding: This work was financially supported by the National Natural Science Foundation of China (22005123); China Postdoctoral Science Foundation (2020M670483), Jiangsu Postdoctoral Science Foundation (2021K382C); the Doctoral Research Foundation of Weifang University (2022BS11, 2022BS09).

Institutional Review Board Statement: Not applicable.

Informed Consent Statement: Not applicable.

Data Availability Statement: Data will be made available on request.

Conflicts of Interest: The authors declare no conflict of interest.

References

1. Li, Y.; Bai, X.; Yuan, D.; Yu, C.; San, X.; Guo, Y.; Zhang, L.; Ye, J. Cu-based high-entropy two-dimensional oxide as stable and active photothermal catalyst. *Nat. Commun.* **2023**, *14*, 3171. [CrossRef]
2. Fan, M.; Jimenez, J.D.; Shirodkar, S.N.; Wu, J.; Chen, S.; Song, L.; Royko, M.M.; Zhang, J.; Guo, H.; Cui, J.; et al. Atomic Ru Immobilized on Porous h-BN through Simple Vacuum Filtration for Highly Active and Selective CO_2 Methanation. *ACS Catal.* **2019**, *9*, 10077–10086. [CrossRef]
3. Yin, H.; Dong, F.; Wang, D.; Li, J. Coupling Cu Single Atoms and Phase Junction for Photocatalytic CO_2 Reduction with 100% CO Selectivity. *ACS Catal.* **2022**, *12*, 14096–14105. [CrossRef]
4. Wang, Z.; Yang, J.; Cao, J.; Chen, W.; Wang, G.; Liao, F.; Zhou, X.; Zhou, F.; Li, R.; Yu, Z.Q.; et al. Room-Temperature Synthesis of Single Iron Site by Electrofiltration for Photoreduction of CO_2 into Tunable Syngas. *ACS Nano* **2020**, *14*, 6164–6172. [CrossRef]
5. Yang, T.; Mao, X.; Zhang, Y.; Wu, X.; Wang, L.; Chu, M.; Pao, C.-W.; Yang, S.; Xu, Y.; Huang, X. Coordination tailoring of Cu single sites on C_3N_4 realizes selective CO_2 hydrogenation at low temperature. *Nat. Commun.* **2021**, *12*, 6022. [CrossRef]
6. Zhu, X.; Zong, H.; Pérez, C.J.V.; Miao, H.; Sun, W.; Yuan, Z.; Wang, S.; Zeng, G.; Xu, H.; Jiang, Z.; et al. Supercharged CO_2 Photothermal Catalytic Methanation: High Conversion, Rate, and Selectivity. *Angew. Chem. Int. Ed.* **2023**, *62*, 202218694. [CrossRef]
7. Zhu, X.; Miao, H.; Chen, J.; Zhu, X.; Yi, J.; Mo, Z.; Li, H.; Zheng, Z.; Huang, B.; Xu, H. Facet-dependent CdS/Bi_4TaO_8Cl Z-scheme heterojunction for enhanced photocatalytic tetracycline hydrochloride degradation and the carrier separation mechanism study via single-particle spectroscopy. *Inorg. Chem. Front.* **2022**, *9*, 2252–2263. [CrossRef]
8. Ju, L.; Tang, X.; Zhang, Y.; Li, X.; Cui, X.; Yang, G. Single Selenium Atomic Vacancy Enabled Efficient Visible-Light-Response Photocatalytic NO Reduction to NH_3 on Janus WSSe Monolayer. *Molecules* **2023**, *28*, 2959. [CrossRef]
9. Saleh, H.M.; Hassan, A.I. 4—Biological conversion of lignocellulosic waste in the renewable energy. In *Advanced Technology for the Conversion of Waste into Fuels and Chemicals*; Khan, A., Jawaid, M., Pizzi, A., Azum, N., Asiri, A., Isa, I., Eds.; Woodhead Publishing: Cambridge, UK, 2021; pp. 99–115.
10. Jia, J.; Qian, C.; Dong, Y.; Li, Y.F.; Wang, H.; Ghoussoub, M.; Butler, K.T.; Walsh, A.; Ozin, G.A. Heterogeneous catalytic hydrogenation of CO_2 by metal oxides defect engineering—Perfecting imperfection. *Chem. Soc. Rev.* **2017**, *46*, 4631–4644. [CrossRef]

11. Liu, H.; Meng, X.; Dao, T.D.; Zhang, H.; Li, P.; Chang, K.; Wang, T.; Li, M.; Nagao, T.; Ye, J. Conversion of Carbon Dioxide by Methane Reforming under Visible-Light Irradiation: Surface-Plasmon-Mediated Nonpolar Molecule Activation. *Angew. Chem. Int. Ed.* **2015**, *54*, 11545–11549. [CrossRef]
12. Yan, T.; Wang, L.; Liang, Y.; Makaremi, M.; Wood, T.E.; Dai, Y.; Huang, B.; Jelle, A.A.; Dong, Y.; Ozin, G.A. Polymorph selection towards photocatalytic gaseous CO_2 hydrogenation. *Nat. Commun.* **2019**, *10*, 2521. [CrossRef]
13. Hoch, L.B.; He, L.; Qiao, Q.; Liao, K.; Reyes, L.M.; Zhu, Y.; Ozin, G.A. Effect of Precursor Selection on the Photocatalytic Performance of Indium Oxide Nanomaterials for Gas-Phase CO_2 Reduction. *Chem. Mater.* **2016**, *28*, 4160–4168. [CrossRef]
14. Wang, J.; Heil, T.; Zhu, B.; Tung, C.W.; Yu, J.; Chen, H.M.; Antonietti, M.; Cao, S. A Single Cu-Center Containing Enzyme-Mimic Enabling Full Photosynthesis under CO_2 Reduction. *ACS Nano* **2020**, *14*, 8584–8593. [CrossRef] [PubMed]
15. Wang, B.; Cai, H.; Shen, S. Single Metal Atom Photocatalysis. *Small Methods* **2019**, *3*, 1800447. [CrossRef]
16. Saleh, H.M.; Hassan, A.I. Green Conversion of Carbon Dioxide and Sustainable Fuel Synthesis. *Fire* **2023**, *6*, 128. [CrossRef]
17. Ju, L.; Tang, X.; Li, J.; Dong, H.; Yang, S.; Gao, Y.; Liu, W. Armchair Janus WSSe Nanotube Designed with Selenium Vacancy as a Promising Photocatalyst for CO_2 Reduction. *Molecules* **2023**, *28*, 4602. [CrossRef]
18. Francke, R.; Schille, B.; Roemelt, M. Homogeneously Catalyzed Electroreduction of Carbon Dioxide-Methods, Mechanisms, and Catalysts. *Chem. Rev.* **2018**, *118*, 4631–4701. [CrossRef]
19. Wang, H.; Jia, J.; Song, P.; Wang, Q.; Li, D.; Min, S.; Qian, C.; Wang, L.; Li, Y.F.; Ma, C.; et al. Efficient Electrocatalytic Reduction of CO_2 by Nitrogen-Doped Nanoporous Carbon Carbon Nanotube Membranes—A Step Towards the Electrochemical CO_2 Refinery. *Angew. Chem. Int. Ed.* **2017**, *56*, 7847–7852. [CrossRef]
20. Gao, S.; Lin, Y.; Jiao, X.; Sun, Y.; Luo, Q.; Zhang, W.; Li, D.; Yang, J.; Xie, Y. Partially oxidized atomic cobalt layers for carbon dioxide electroreduction to liquid fuel. *Nature* **2016**, *529*, 68–71. [CrossRef]
21. Gao, S.; Sun, Z.; Liu, W.; Jiao, X.; Zu, X.; Hu, Q.; Sun, Y.; Yao, T.; Zhang, W.; Wei, S.; et al. Atomic layer confined vacancies for atomic-level insights into carbon dioxide electroreduction. *Nat. Commun.* **2017**, *8*, 14503. [CrossRef]
22. Chen, P.; Lei, B.; Dong, X.; Wang, H.; Sheng, J.; Cui, W.; Li, J.; Sun, Y.; Wang, Z.; Dong, F. Rare-Earth Single-Atom La-N Charge-Transfer Bridge on Carbon Nitride for Highly Efficient and Selective Photocatalytic CO_2 Reduction. *ACS Nano* **2020**, *14*, 15841–15852. [CrossRef]
23. Huang, P.; Huang, J.; Pantovich, S.A.; Carl, A.D.; Fenton, T.G.; Caputo, C.A.; Grimm, R.L.; Frenkel, A.I.; Li, G. Selective CO_2 Reduction Catalyzed by Single Cobalt Sites on Carbon Nitride under Visible-Light Irradiation. *J. Am. Chem. Soc.* **2018**, *140*, 16042–16047. [CrossRef]
24. Cao, Y.; Guo, L.; Dan, M.; Doronkin, D.E.; Han, C.; Rao, Z.; Liu, Y.; Meng, J.; Huang, Z.; Zheng, K.; et al. Modulating electron density of vacancy site by single Au atom for effective CO_2 photoreduction. *Nat. Commun.* **2021**, *12*, 1675. [CrossRef] [PubMed]
25. Chen, J.; Zhu, X.; Jiang, Z.; Zhang, W.; Ji, H.; Zhu, X.; Song, Y.; Mo, Z.; Li, H.; Xu, H. Construction of brown mesoporous carbon nitride with a wide spectral response for high performance photocatalytic H_2 evolution. *Inorg. Chem. Front.* **2022**, *9*, 103–110. [CrossRef]
26. Zhang, Z.; Li, D.; Chu, Y.; Chang, L.; Xu, J. Space-Confined Growth of Cs_2CuBr_4 Perovskite Nanodots in Mesoporous CeO_2 for Photocatalytic CO_2 Reduction: Structure Regulation and Built-in Electric Field Construction. *J. Phys. Chem. Lett.* **2023**, *14*, 5249–5259. [CrossRef] [PubMed]
27. Jiang, Y.; Zhou, R.; Zhang, Z.; Dong, Z.; Xu, J. Boosted charge transfer and CO_2 photoreduction by construction of S-scheme heterojunctions between $Cs_2AgBiBr_6$ nanosheets and two-dimensional metal–organic frameworks. *J. Mater. Chem. C* **2023**, *11*, 2540–2551. [CrossRef]
28. Zhang, Z.; Li, D.; Dong, Z.; Jiang, Y.; Li, X.; Chu, Y.; Xu, J. Lead-Free $Cs_2AgBiBr_6$ Nanocrystals Confined in MCM-48 Mesoporous Molecular Sieve for Efficient Photocatalytic CO_2 Reduction. *Sol. RRL* **2023**, *7*, 2300038. [CrossRef]
29. Shehzad, N.; Tahir, M.; Johari, K.; Murugesan, T.; Hussain, M. A critical review on TiO_2 based photocatalytic CO_2 reduction system: Strategies to improve efficiency. *J. CO2 Util.* **2018**, *26*, 98–122. [CrossRef]
30. Ye, S.; Wang, R.; Wu, M.-Z.; Yuan, Y.-P. A review on g-C_3N_4 for photocatalytic water splitting and CO_2 reduction. *Appl. Surf. Sci.* **2015**, *358*, 15–27. [CrossRef]
31. Zhao, M.; Huang, Y.; Peng, Y.; Huang, Z.; Ma, Q.; Zhang, H. Two-dimensional metal-organic framework nanosheets: Synthesis and applications. *Chem. Soc. Rev.* **2018**, *47*, 6267–6295. [CrossRef]
32. Wu, J.; Li, X.; Shi, W.; Ling, P.; Sun, Y.; Jiao, X.; Gao, S.; Liang, L.; Xu, J.; Yan, W.; et al. Efficient Visible-Light-Driven CO_2 Reduction Mediated by Defect-Engineered BiOBr Atomic Layers. *Angew. Chem. Int. Ed. Engl.* **2018**, *57*, 8719–8723. [CrossRef] [PubMed]
33. Ashley, A.E.; Thompson, A.L.; O'Hare, D. Non-metal-mediated homogeneous hydrogenation of CO_2 to CH_3OH. *Angew. Chem. Int. Ed. Engl.* **2009**, *48*, 9839–9843. [CrossRef] [PubMed]
34. Kattel, S.; Ramírez, P.J.; Chen, J.G.; Rodriguez, J.A.; Liu, P. Active sites for CO_2 hydrogenation to methanol on Cu/ZnO catalysts. *Science* **2017**, *355*, 1296–1299. [CrossRef] [PubMed]
35. Grabowski, R.; Słoczyński, J.; Śliwa, M.; Mucha, D.; Socha, R.P.; Lachowska, M.; Skrzypek, J. Influence of Polymorphic ZrO_2 Phases and the Silver Electronic State on the Activity of Ag/ZrO_2 Catalysts in the Hydrogenation of CO_2 to Methanol. *ACS Catal.* **2011**, *1*, 266–278. [CrossRef]
36. Wang, W.; Wang, S.; Ma, X.; Gong, J. Recent advances in catalytic hydrogenation of carbon dioxide. *Chem. Soc. Rev.* **2011**, *40*, 3703–3727. [CrossRef]

37. Kondratenko, E.V.; Mul, G.; Baltrusaitis, J.; Larrazábal, G.O.; Pérez-Ramírez, J. Status and perspectives of CO_2 conversion into fuels and chemicals by catalytic, photocatalytic and electrocatalytic processes. *Energy Environ. Sci.* **2013**, *6*, 3112–3135. [CrossRef]
38. Tountas, A.A.; Peng, X.; Tavasoli, A.V.; Duchesne, P.N.; Dingle, T.L.; Dong, Y.; Hurtado, L.; Mohan, A.; Sun, W.; Ulmer, U.; et al. Towards Solar Methanol Past, Present, and Future. *Adv. Sci.* **2019**, *6*, 1801903. [CrossRef]
39. Meng, X.; Wang, T.; Liu, L.; Ouyang, S.; Li, P.; Hu, H.; Kako, T.; Iwai, H.; Tanaka, A.; Ye, J. Photothermal conversion of CO_2 into CH_4 with H_2 over Group VIII nanocatalysts: An alternative approach for solar fuel production. *Angew. Chem. Int. Ed. Engl.* **2014**, *53*, 11478–11482. [CrossRef]
40. Qian, C.; Sun, W.; Hung, D.L.H.; Qiu, C.; Makaremi, M.; Kumar, S.G.H.; Wan, L.; Ghoussoub, M.; Wood, T.E.; Xia, M.; et al. Catalytic CO_2 reduction by palladium-decorated silicon–hydride nanosheets. *Nat. Catal.* **2019**, *2*, 46–54. [CrossRef]
41. Yang, P.P.; Zhang, X.L.; Gao, F.Y.; Zheng, Y.R.; Niu, Z.Z.; Yu, X.; Liu, R.; Wu, Z.Z.; Qin, S.; Chi, L.P.; et al. Protecting Copper Oxidation State via Intermediate Confinement for Selective CO_2 Electroreduction to C_{2+} Fuels. *J. Am. Chem. Soc.* **2020**, *142*, 6400–6408. [CrossRef]
42. Luc, W.; Fu, X.; Shi, J.; Lv, J.-J.; Jouny, M.; Ko, B.H.; Xu, Y.; Tu, Q.; Hu, X.; Wu, J.; et al. Two-dimensional copper nanosheets for electrochemical reduction of carbon monoxide to acetate. *Nat. Catal.* **2019**, *2*, 423–430. [CrossRef]
43. Chen, C.; Li, Y.; Yu, S.; Louisia, S.; Jin, J.; Li, M.; Ross, M.B.; Yang, P. Cu-Ag Tandem Catalysts for High-Rate CO_2 Electrolysis toward Multicarbons. *Joule* **2020**, *4*, 1688–1699. [CrossRef]
44. Voiry, D.; Chhowalla, M.; Gogotsi, Y.; Kotov, N.A.; Li, Y.; Penner, R.M.; Schaak, R.E.; Weiss, P.S. Best Practices for Reporting Electrocatalytic Performance of Nanomaterials. *ACS Nano* **2018**, *12*, 9635–9638. [CrossRef] [PubMed]
45. Kong, S.; Lv, X.; Wang, X.; Liu, Z.; Li, Z.; Jia, B.; Sun, D.; Yang, C.; Liu, L.; Guan, A.; et al. Delocalization state-induced selective bond breaking for efficient methanol electrosynthesis from CO_2. *Nat. Catal.* **2023**, *6*, 6–15. [CrossRef]
46. Tian, F.; Zhang, H.; Liu, S.; Wu, T.; Yu, J.; Wang, D.; Jin, X.; Peng, C. Visible-light-driven CO_2 reduction to ethylene on CdS: Enabled by structural relaxation-induced intermediate dimerization and enhanced by ZIF-8 coating. *Appl. Catal. B-Environ.* **2021**, *285*, 119834. [CrossRef]
47. Wang, B.; Zhang, W.; Liu, G.; Chen, H.; Weng, Y.X.; Li, H.; Chu, P.K.; Xia, J. Excited Electron-Rich $Bi^{(3-x)+}$ Sites: A Quantum Well-Like Structure for Highly-Promoted Selective Photocatalytic CO_2 Reduction Performance. *Adv. Funct. Mater.* **2022**, *32*, 2202885. [CrossRef]
48. Yang, J.; Hao, J.; Xu, S.; Wang, Q.; Dai, J.; Zhang, A.; Pang, X. $InVO_4$/beta-$AgVO_3$ Nanocomposite as a Direct Z-Scheme Photocatalyst toward Efficient and Selective Visible-Light-Driven CO_2 Reduction. *ACS Appl. Mater. Interfaces* **2019**, *11*, 32025–32037. [CrossRef] [PubMed]
49. Wang, S.; Jiang, B.; Henzie, J.; Xu, F.; Liu, C.; Meng, X.; Zou, S.; Song, H.; Pan, Y.; Li, H.; et al. Designing reliable and accurate isotope-tracer experiments for CO_2 photoreduction. *Nat. Commun.* **2023**, *14*, 2534. [CrossRef]
50. Wang, S.; Guan, B.Y.; Lu, Y.; Lou, X.W.D. Formation of Hierarchical In_2S_3–$CdIn_2S_4$ Heterostructured Nanotubes for Efficient and Stable Visible Light CO_2 Reduction. *J. Am. Chem. Soc.* **2017**, *139*, 17305–17308. [CrossRef]
51. Putta Rangappa, A.; Praveen Kumar, D.; Do, K.H.; Wang, J.; Zhang, Y.; Kim, T.K. Synthesis of Pore-Wall-Modified Stable COF/TiO_2 Heterostructures via Site-Specific Nucleation for an Enhanced Photoreduction of Carbon Dioxide. *Adv. Sci.* **2023**, *10*, 2300073. [CrossRef]
52. Singh, P.; Yadav, R.K.; Singh, C.; Chaubey, S.; Singh, S.; Singh, A.P.; Baeg, J.-O.; Kim, T.W.; Gulzhian, D. Photocatalytic activity of ultrathin 2DPNs for enzymatically generating formic acid from CO_2 and C–S/C–N bond formation. *Sustain. Energy Fuels* **2022**, *6*, 2223–2232. [CrossRef]
53. Kumar, S.; Yadav, R.K.; Choi, S.Y.; Singh, P.; Kim, T.W. An efficient polydopamine modified sulphur doped GCN photocatalyst for generation of HCOOH from CO_2 under sun ray irradiation. *J. Photo. Photobio. A* **2023**, *439*, 114591. [CrossRef]
54. Yang, Z.; Qi, Y.; Wang, F.; Han, Z.; Jiang, Y.; Han, H.; Liu, J.; Zhang, X.; Ong, W.J. State-of-the-art advancements in photo-assisted CO_2 hydrogenation: Recent progress in catalyst development and reaction mechanisms. *J. Mater. Chem. A* **2020**, *8*, 24868–24894. [CrossRef]
55. Fan, Y.; Ma, X.; Liu, X.; Wang, J.; Ai, H.; Zhao, M. Theoretical Design of an InSe/GaTe vdW Heterobilayer: A Potential Visible-Light Photocatalyst for Water Splitting. *J. Phys. Chem. C* **2018**, *122*, 27803–27810. [CrossRef]
56. Guiglion, P.; Butchosa, C.; Zwijnenburg, M.A. Polymer Photocatalysts for Water Splitting: Insights from Computational Modeling. *Macromol. Chem. Phys.* **2016**, *217*, 344–353. [CrossRef]
57. Shaikh, A.R.; Posada-Pérez, S.; Brotons-Rufes, A.; Pajski, J.J.; Vajiha; Kumar, G.; Mateen, A.; Poater, A.; Solà, M.; Chawla, M.; et al. Selective absorption of H_2S and CO_2 by azole based protic ionic liquids: A combined density functional theory and molecular dynamics study. *J. Mol. Liq.* **2022**, *367*, 120558. [CrossRef]
58. Posada-Pérez, S.; Vidal-López, A.; Solà, M.; Poater, A. 2D carbon nitride as a support with single Cu, Ag, and Au atoms for carbon dioxide reduction reaction. *Phys. Chem. Chem. Phys.* **2023**, *25*, 8574–8582. [CrossRef]
59. Varvoutis, G.; Lykaki, M.; Marnellos, G.E.; Konsolakis, M. Recent Advances on Fine-Tuning Engineering Strategies of CeO_2-Based Nanostructured Catalysts Exemplified by CO_2 Hydrogenation Processes. *Catalysts* **2023**, *13*, 275. [CrossRef]
60. Leung, C.-F.; Ho, P.-Y. Molecular Catalysis for Utilizing CO_2 in Fuel Electro-Generation and in Chemical Feedstock. *Catalysts* **2019**, *9*, 760. [CrossRef]
61. Posada-Pérez, S.; Solà, M.; Poater, A. Carbon Dioxide Conversion on Supported Metal Nanoparticles: A Brief Review. *Catalysts* **2023**, *13*, 305. [CrossRef]

62. Qiu, L.-Q.; Yao, X.; Zhang, Y.-K.; Li, H.-R.; He, L.-N. Advancements and Challenges in Reductive Conversion of Carbon Dioxide via Thermo-/Photocatalysis. *J. Org. Chem.* **2023**, *88*, 4942–4964. [CrossRef] [PubMed]
63. Meng, C.; Zhao, G.; Shi, X.-R.; Chen, P.; Liu, Y.; Lu, Y. Oxygen-deficient metal oxides supported nano-intermetallic InNi$_3$C$_{0.5}$ toward efficient CO$_2$ hydrogenation to methanol. *Sci. Adv.* **2021**, *7*, eabi6012. [CrossRef] [PubMed]
64. Hu, J.; Yu, L.; Deng, J.; Wang, Y.; Cheng, K.; Ma, C.; Zhang, Q.; Wen, W.; Yu, S.; Pan, Y.; et al. Sulfur vacancy-rich MoS$_2$ as a catalyst for the hydrogenation of CO$_2$ to methanol. *Nat. Catal.* **2021**, *4*, 242–250. [CrossRef]
65. Marxer, D.; Furler, P.; Takacs, M.; Steinfeld, A. Solar thermochemical splitting of CO$_2$ into separate streams of CO and O$_2$ with high selectivity, stability, conversion, and efficiency. *Energy Environ. Sci.* **2017**, *10*, 1142–1149. [CrossRef]
66. Heidlage, M.G.; Kezar, E.A.; Snow, K.C.; Pfromm, P.H. Thermochemical Synthesis of Ammonia and Syngas from Natural Gas at Atmospheric Pressure. *Ind. Eng. Chem. Res.* **2017**, *56*, 14014–14024. [CrossRef]
67. Gokon, N.; Yamawaki, Y.; Nakazawa, D.; Kodama, T. Ni/MgO–Al$_2$O$_3$ and Ni–Mg–O catalyzed SiC foam absorbers for high temperature solar reforming of methane. *Int. J. Hydrogen Energy* **2010**, *35*, 7441–7453. [CrossRef]
68. Tou, M.; Michalsky, R.; Steinfeld, A. Solar-Driven Thermochemical Splitting of CO$_2$ and In Situ Separation of CO and O$_2$ across a Ceria Redox Membrane Reactor. *Joule* **2017**, *1*, 146–154. [CrossRef]
69. Chueh, W.C.; Haile, S.M. Ceria as a thermochemical reaction medium for selectively generating syngas or methane from H$_2$O and CO$_2$. *ChemSusChem* **2009**, *2*, 735–739. [CrossRef]
70. Feng, K.; Wang, S.; Zhang, D.; Wang, L.; Yu, Y.; Feng, K.; Li, Z.; Zhu, Z.; Li, C.; Cai, M.; et al. Cobalt Plasmonic Superstructures Enable Almost 100% Broadband Photon Efficient CO$_2$ Photocatalysis. *Adv. Mater.* **2020**, *32*, 2000014. [CrossRef] [PubMed]
71. Linic, S.; Christopher, P.; Ingram, D.B. Plasmonic-metal nanostructures for efficient conversion of solar to chemical energy. *Nat. Mater.* **2011**, *10*, 911–921. [CrossRef] [PubMed]
72. Choi, K.M.; Kim, D.; Rungtaweevoranit, B.; Trickett, C.A.; Barmanbek, J.T.D.; Alshammari, A.S.; Yang, P.; Yaghi, O.M. Plasmon-Enhanced Photocatalytic CO$_2$ Conversion within Metal–Organic Frameworks under Visible Light. *J. Am. Chem. Soc.* **2017**, *139*, 356–362. [CrossRef] [PubMed]
73. Xu, C.; Huang, W.; Li, Z.; Deng, B.; Zhang, Y.; Ni, M.; Cen, K. Photothermal Coupling Factor Achieving CO$_2$ Reduction Based on Palladium-Nanoparticle-Loaded TiO$_2$. *ACS Catal.* **2018**, *8*, 6582–6593. [CrossRef]
74. Jia, J.; O'Brien, P.G.; He, L.; Qiao, Q.; Fei, T.; Reyes, L.M.; Burrow, T.E.; Dong, Y.; Liao, K.; Varela, M.; et al. Visible and Near-Infrared Photothermal Catalyzed Hydrogenation of Gaseous CO$_2$ over Nanostructured Pd@Nb$_2$O$_5$. *Adv. Sci.* **2016**, *3*, 1600189. [CrossRef]
75. Li, Y.F.; Soheilnia, N.; Greiner, M.; Ulmer, U.; Wood, T.; Jelle, A.A.; Dong, Y.; Yin Wong, A.P.; Jia, J.; Ozin, G.A. Pd@HyWO$_{3-x}$ Nanowires Efficiently Catalyze the CO$_2$ Heterogeneous Reduction Reaction with a Pronounced Light Effect. *ACS Appl. Mater. Interfaces* **2019**, *11*, 5610–5615. [CrossRef] [PubMed]
76. Ren, J.; Ouyang, S.; Xu, H.; Meng, X.; Wang, T.; Wang, D.; Ye, J. Targeting Activation of CO$_2$ and H$_2$ over Ru-Loaded Ultrathin Layered Double Hydroxides to Achieve Efficient Photothermal CO$_2$ Methanation in Flow-Type System. *Adv. Energy Mater.* **2017**, *7*, 1601657. [CrossRef]
77. Hong, J.; Zhang, W.; Wang, Y.; Zhou, T.; Xu, R. Photocatalytic Reduction of Carbon Dioxide over Self-Assembled Carbon Nitride and Layered Double Hydroxide: The Role of Carbon Dioxide Enrichment. *ChemCatChem* **2014**, *6*, 2315–2321. [CrossRef]
78. Ahmed, N.; Shibata, Y.; Taniguchi, T.; Izumi, Y. Photocatalytic conversion of carbon dioxide into methanol using zinc–copper–M(III) (M=aluminum, gallium) layered double hydroxides. *J. Catal.* **2011**, *279*, 123–135. [CrossRef]
79. Morikawa, M.; Ahmed, N.; Yoshida, Y.; Izumi, Y. Photoconversion of carbon dioxide in zinc–copper–gallium layered double hydroxides: The kinetics to hydrogen carbonate and further to CO/methanol. *Appl. Catal. B-Environ.* **2014**, *144*, 561–569. [CrossRef]
80. Lin, L.; Wang, K.; Yang, K.; Chen, X.; Fu, X.; Dai, W. The visible-light-assisted thermocatalytic methanation of CO$_2$ over Ru/TiO$_{(2-x)}$N$_x$. *Appl. Catal. B-Environ.* **2017**, *204*, 440–455. [CrossRef]
81. O'Brien, P.G.; Sandhel, A.; Wood, T.E.; Jelle, A.A.; Hoch, L.B.; Perovic, D.D.; Mims, C.A.; Ozin, G.A. Photomethanation of Gaseous CO$_2$ over Ru/Silicon Nanowire Catalysts with Visible and Near-Infrared Photons. *Adv. Sci.* **2014**, *1*, 1400001. [CrossRef]
82. Jelle, A.A.; Ghuman, K.K.; O'Brien, P.G.; Hmadeh, M.; Sandhel, A.; Perovic, D.D.; Singh, C.V.; Mims, C.A.; Ozin, G.A. Highly Efficient Ambient Temperature CO$_2$ Photomethanation Catalyzed by Nanostructured RuO$_2$ on Silicon Photonic Crystal Support. *Adv. Energy Mater.* **2018**, *8*, 1702277. [CrossRef]
83. Zhang, X.; Li, X.; Zhang, D.; Su, N.Q.; Yang, W.; Everitt, H.O.; Liu, J. Product selectivity in plasmonic photocatalysis for carbon dioxide hydrogenation. *Nat. Commun.* **2017**, *8*, 14542. [CrossRef] [PubMed]
84. Liu, H.; Song, H.; Zhou, W.; Meng, X.; Ye, J. A Promising Application of Optical Hexagonal TaN in Photocatalytic Reactions. *Angew. Chem. Int. Ed.* **2018**, *57*, 16781–16784. [CrossRef] [PubMed]

Disclaimer/Publisher's Note: The statements, opinions and data contained in all publications are solely those of the individual author(s) and contributor(s) and not of MDPI and/or the editor(s). MDPI and/or the editor(s) disclaim responsibility for any injury to people or property resulting from any ideas, methods, instructions or products referred to in the content.

Review

A Review on Cu₂O-Based Composites in Photocatalysis: Synthesis, Modification, and Applications

Qian Su, Cheng Zuo *, Meifang Liu * and Xishi Tai *

College of Chemistry & Chemical and Environmental Engineering, Weifang University, Weifang 261061, China; sqian316@wfu.edu.cn
* Correspondence: chengzuo@wfu.edu.cn (C.Z.); liumf@iccas.ac.cn (M.L.); taixs@wfu.edu.cn (X.T.)

Abstract: Photocatalysis technology has the advantages of being green, clean, and environmentally friendly, and has been widely used in CO_2 reduction, hydrolytic hydrogen production, and the degradation of pollutants in water. Cu_2O has the advantages of abundant reserves, a low cost, and environmental friendliness. Based on the narrow bandgap and strong visible light absorption ability of Cu_2O, Cu_2O-based composite materials show infinite development potential in photocatalysis. However, in practical large-scale applications, Cu_2O-based composites still pose some urgent problems that need to be solved, such as the high composite rate of photogenerated carriers, and poor photocatalytic activity. This paper introduces a series of Cu_2O-based composites, based on recent reports, including pure Cu_2O and Cu_2O hybrid materials. The modification strategies of photocatalysts, critical physical and chemical parameters of photocatalytic reactions, and the mechanism for the synergistic improvement of photocatalytic performance are investigated and explored. In addition, the application and photocatalytic performance of Cu_2O-based photocatalysts in CO_2 photoreduction, hydrogen production, and water pollution treatment are discussed and evaluated. Finally, the current challenges and development prospects are pointed out, to provide guidance in applying Cu_2O-based catalysts in renewable energy utilization and environmental protection.

Keywords: heterojunction; photocatalysis; synthesis; modification; application

Citation: Su, Q.; Zuo, C.; Liu, M.; Tai, X. A Review on Cu₂O-Based Composites in Photocatalysis: Synthesis, Modification, and Applications. *Molecules* **2023**, *28*, 5576. https://doi.org/10.3390/molecules28145576

Academic Editor: Lin Ju

Received: 10 July 2023
Revised: 20 July 2023
Accepted: 21 July 2023
Published: 22 July 2023

Copyright: © 2023 by the authors. Licensee MDPI, Basel, Switzerland. This article is an open access article distributed under the terms and conditions of the Creative Commons Attribution (CC BY) license (https://creativecommons.org/licenses/by/4.0/).

1. Introduction

With the development of industrialization, the use of fossil fuels in industry has caused many problems, such as carbon dioxide emissions causing global warming, water pollution, and the destruction of surrounding biological habitats. Energy shortages and environmental pollution pose a serious threat to the development of industry and agriculture, and have become hot topics that need to be addressed [1,2]. Photocatalytic technology utilizes semiconductor materials to achieve the photoreduction of CO_2, the photocatalytic decomposition of water, and the degradation of pollutants, and has the advantages of a low cost, simple operation, and no secondary pollution [3–5]. Photocatalysis is a green technology that fully utilizes solar energy, and is considered one of the most feasible and promising methods to solve environmental and energy problems.

Since Fukushima et al. [6] discovered in 1967 that TiO_2 can decompose water to produce hydrogen under light, tremendous progress has been made in photocatalytic technology. Due to its stable structure, high efficiency, low cost, nontoxicity, and high optical stability, TiO_2 has become widely studied in the past few decades [7,8]. However, TiO_2 can only absorb 3–5% of total ultraviolet light, so its utilization of sunlight is not high, significantly limiting its practical application under sunlight [9,10]. To effectively utilize the maximum proportion of visible light covering the solar spectrum ($\lambda > 400$ nm), in addition to modifying TiO_2, researchers have studied a series of novel photocatalysts with a visible light response, such as simple oxides (ZnO [11] and Cu_2O [12]), sulfides (CdS [13] and MoS_2 [14]), Bi-based materials (Bi_2WO_6 [15] and $BiVO_4$ [16]), and nitrides (C_3N_4 [17]).

Compared with other semiconductors, copper(I) oxide (Cu_2O) has the advantages of nontoxicity, a favorable environmental acceptability, low cost, and high activity. It has been widely used in solar cells [18], carbon monoxide oxidation [19], photocatalysts [20], electrocatalysts [21], and sensors [22]. As a p-type semiconductor, Cu_2O has a bandgap width of 2.17 eV, and a broad response range to the solar spectrum. A Cu_2O material is currently one of the most promising visible light photocatalysts, and has become a research hotspot in photocatalysis. Li et al. [23] researched cubic c-Cu_2O with the main exposure surface of (100), and tested its photocatalytic degradation performance on methyl orange (MO). It could completely decompose MO in an aqueous solution within 80 min under visible light, and almost remained unchanged in five consecutive cycles, showing a satisfactory stability. However, the application prospects of Cu_2O in photocatalysis have been limited due to its poor stability, susceptibility to photocorrosion, and low quantum yield. To further improve the photocatalytic performance and stability of Cu_2O, researchers have focused on a series of studies on morphology control, heteroatom doping, and the construction of semiconductor heterojunctions. With the increasing maturity in preparation and detection methods, the research on improving the photocatalytic activity of Cu_2O has become more in-depth and diversified.

Unlike other reports in the literature [24–26], this paper reviews the preparation methods and applications of different Cu_2O-based composites reported in recent years. By analyzing the roles of the different components in the photocatalytic process, we explain the reasons behind the improvement of photocatalytic performance, and point out the future direction that the industrial application of Cu_2O-based composites could take.

According to data from the Web of Science platform by Clarivate Analytics (Figure 1), research on the subject of the photocatalysis of Cu_2O and Cu_2O-based materials is increasing year by year, indicating that Cu_2O-based materials are becoming ideal candidates for a variety of energy and environmental photocatalysis applications. Based on the research direction of Cu_2O photocatalysts, in this paper, the main preparation methods are introduced, including their merits and demerits, and the current main research focus and progress are reviewed. The research ideas and framework of this review are shown in Figure 2. The frequently used methods for improving their photocatalytic performance are reviewed, including morphology and size improvement, doping, metal loading, and semiconductor hybridization. In addition, several representative Cu_2O-based composite photocatalysts are introduced, including metal/Cu_2O composite, Cu_2O semiconductor composite, Cu_2O/carbon material composite, and ternary composite photocatalysts. Their applications in the photocatalytic reduction of CO_2, photocatalytic hydrolysis of hydrogen, photocatalytic degradation of pollutants, and photocatalytic reduction of metal atoms are discussed through the combination of the experimental numbers and reaction principles. Finally, the future development of Cu_2O-based composite photocatalysts is considered.

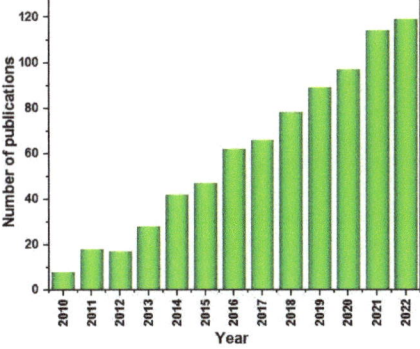

Figure 1. The annual number of publications using "Cu_2O" as a topic keyword since 2010 (data taken from Web of Science on 1 January 2023).

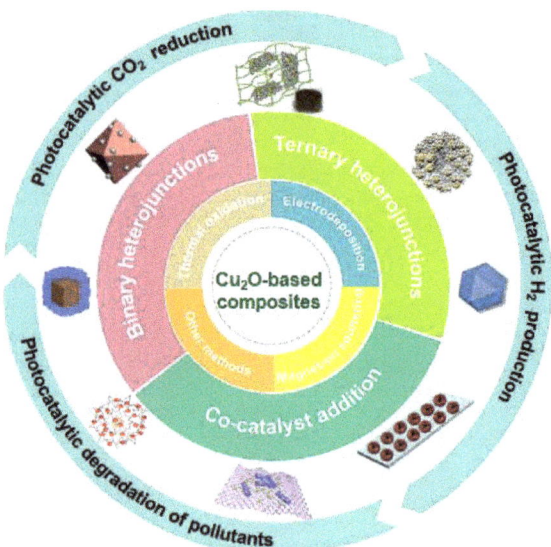

Figure 2. The research framework and basis of thinking for this review.

2. Synthetic Methods

The low utilization of visible light by single CuO materials, and the easy complexation of electron–hole pairs generated by CuO under photoexcitation limit the application of CuO materials in photocatalysis. There is currently more research on doping Cu_2O with high Li or Na metal concentrations. The bonding network of the off-domain two- and three-electron centers is disrupted, effectively localizing the electrons in the limited space.

2.1. Preparation Methods

Different preparation methods to prepare effective Cu_2O materials for photocatalytic experiments have been reported in the literature. The thermal oxidation of metals is a widely used method for synthesizing high-quality oxides. The final desired thickness of the Cu_2O layer is prepared based on the oxidation of the high-purity copper foil. The temperature range is between 1000 and 1500 °C under a pure oxygen atmosphere or mixed gas atmosphere (e.g., $Ar + O_2$). The obtained Cu_2O is polycrystalline, with different grain structures depending on the chosen experimental conditions. In general, a mixture of CuO and Cu_2O appears in copper foil at the end of oxidation. The Cu_2O appears first, at the beginning of the oxidation process, while the CuO takes a long time to appear in the oxidation process. Electrodeposition is one of the methods for the production of high-quality Cu_2O. The advantages of this method are that it is cheap, can efficiently work on different substrates, and allows for adjustments in the material properties and morphology according to the following parameters: the applied potential, current, and temperature, and the pH of the tank solution. The first electrochemical synthesis of Cu_2O was by Stareck [27]. Subsequently, many other scholars developed different synthesis methods, using copper precursors, electrolytes, and electrochemistry. Two types of Cu_2O nanoparticles were successfully synthesized by adjusting the pH of the electrolyte [28]. Firstly, the pH of the electrolyte was adjusted to 10, and a pyramid-shaped p-type Cu_2O crystal was grown on the FTO substrate. Subsequently, the pH was adjusted to 4.9, and an ultra-thin layer of n-Cu_2O deposition product was obtained (Figure 3). The p-Cu_2O nanoparticles and the n-Cu_2O protective layer on the surface formed a p/n heterojunction. The modified p/n-Cu_2O had a bandgap of 2–2.2 eV, and could be excited by visible light with a wavelength less than 600 nm. Its photocurrent response was significantly improved, increasing the charge transfer rate and the stability of the catalyst. Thus, the modified p/n-Cu$_2$O catalyst exhibited much higher activity than the original p-Cu$_2$O.

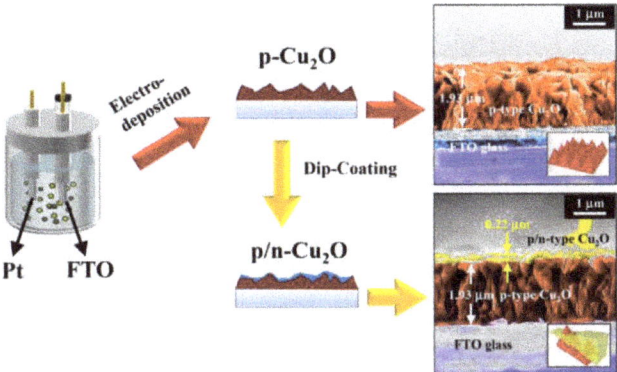

Figure 3. Schematic diagram of the procedure for the electrodeposition of p/n Cu$_2$O on the FTO substrate [28].

Magnetron sputtering is a process that uses high-energy particles to bombard a solid target, so that atoms or molecules sputtered from the surface of the target form a thin film in a specific region. The CuO films prepared using this method exhibit nanometer-sized columnar structures, and the crystallinity, grain size, and film thickness of the Cu$_2$O films can be controlled by varying the sputtering parameters (e.g., the sputtering power, oxygen content, oxygen concentration, sputter deposition time, and annealing temperature). Cu$_2$O–CuO films with an excellent photocatalytic performance have been deposited on glass substrates using RF magnetron sputtering (Figure 4) [29]. It has been observed that with the prolongation of the sputtering deposition time, the size of the Cu$_2$O–CuO nanoparticles has increased from 7 nm to 13 nm, and the thickness of the thin films from 7 nm to 50 nm, resulting in a rougher surface, reduced bandgap, and decreased PL strength. The results indicate that the structure, morphology, and optical and photocatalytic properties of prepared Cu$_2$O–CuO films are strongly dependent on the deposition thickness. Under sunlight exposure, Cu$_2$O–CuO films can completely degrade pollutants (methylene blue and methyl orange) from water within only 60 min.

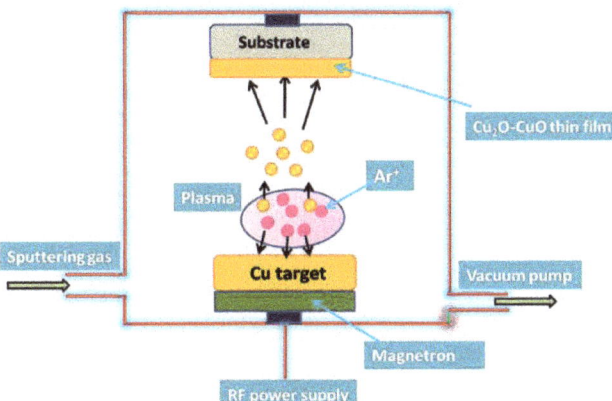

Figure 4. Schematic diagram of a Cu$_2$O–CuO film prepared using the magnetron sputtering method [29]. Copyright 2020, Elsevier.

2.2. Other Methods

In addition to the above methods, different surfactants [30–35] and micelles [36] have been used, mainly to control the morphology of the prepared Cu-based catalyst particles. Cu_2O nanocrystals and nanoarray with cubic [37–39], octahedral [40], and multipod structures [41] have been prepared using these methods. Yang et al. [42] have proposed a metal-induced thermal reduction (MITR) method for the in-situ growth of Cu_2O crystals on a copper substrate. The corresponding scheme is shown in Figure 5, and the operation is divided into two steps: (a) under alkaline conditions, the $Cu(OH)_2$ nanorod array is in-situ grown by impregnating copper foil with a mixed solution of $(NH_4)_2S_2O_8$ and ammonia; and (b) the $Cu(OH)_2$ on copper foil is directly thermally reduced to Cu_2O nanorod array films in a N_2 atmosphere at 500 °C. The average diameter of a nanorod was 400 ± 100 nm, with a length of several micrometers. The method is simple and efficient, and the preparation process has a low energy consumption and is controllable. In addition, the introduction of the substrate metal Cu can significantly reduce the reduction temperature, by changing the Gibbs free energy of the reaction. Surfactant-free synthesis has also been developed to reduce the interference of these surfactants [43–45]. Solvothermal [46,47] and sol–gel [48] methods have also been tested. The wet chemistry route [49,50], thermal evaporation [51,52], chemical vapor deposition [53,54], and hydrothermal route [55–58] are also common methods for synthesizing such semiconductors. In addition, the corresponding properties of Cu_2O-based materials synthesized by different methods are detailed in the Section 4, including their morphology, structure, band gap, and photocatalytic applications.

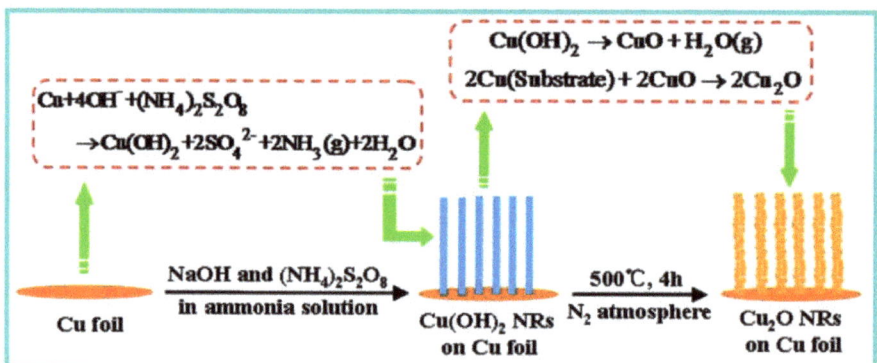

Figure 5. The schematic illustration of the synthesis process of Cu_2O nanorod array films [42]. Copyright 2016, Elsevier.

3. Modification Strategies

Although noble metals have been used in photocatalytic organic waste degradation and CO_2 reduction, their efficiency is still high. However, the cost is also high (e.g., Pt and Au), making them unsuitable for future industrial development. In contrast, Cu_2O is inexpensive to use. It also has excellent CO_2 capture ability and photochemical and structural properties, and shows unlimited development potential in CO_2 reduction. However, the high electron–hole complexation rate and the low optical quantum efficiency limit the application of Cu_2O in photocatalysis. To improve the photocatalytic efficiency of Cu_2O, the structure of Cu_2O needs to be modified. The modified structures are mainly divided into binary and ternary Cu_2O heterostructure structures, and the addition of co-catalysts, in this section.

3.1. Binary Cu_2O-Based Heterojunctions

3.1.1. Cu_2O/Noble Metal Heterojunction

The Fermi energy level of the noble metal material is relatively low compared to that of the catalyst in the photocatalytic reduction of CO_2, which has a higher work function than that of the catalyst. The mutual contact between the two will form a Schottky barrier at the metal–semiconductor interface, which can effectively inhibit the complexation of photoexcited electron–hole pairs, thus promoting the catalytic process, and improving the catalytic efficiency of the catalyst. The currently synthesized Cu_2O/noble metal composites are Cu_2O/Ag [59], Cu_2O/Au [60], and Cu_2O/Pt [61]. These materials show more than 90% photocatalytic efficiency for modified Cu_2O.

Cu_2O/Au nanostructures have been extensively investigated in recent years. Kuo et al. [62] reported the synthesis of Au@Cu_2O core–shell nanocrystals using a chemical reduction method. The nanocrystals exhibited high activity in degrading methyl orange. Ag is relatively inexpensive, and has a higher electron transfer efficiency than metallic Au. Therefore, Ag/Cu_2O catalysts have been more widely studied. Yang et al. [63] prepared Cu_2O/Ag spherical microstructures by depositing silver nanoparticles on the surface of Cu_2O through the thermal decomposition of silver acetate.

3.1.2. Cu_2O/Graphene (GO) Heterojunction

From amorphous carbon black to crystalline structured natural layered graphite, and from zero-dimensional nanostructured fullerenes to two-dimensional structured graphene, carbon materials have been the most widely used and endlessly promising materials on earth. In recent decades, carbon nanomaterials have attracted much attention. The discovery of graphene self-assembled hydrogels with three-dimensional mesh structures has dramatically enriched the carbon material family, and provided a new growth point for new materials. Due to their unique nanostructure and properties, they have also shown significant scientific significance and experimental results. Thus, they provide a new target and direction when it comes to researching carbon-based materials. Graphene has been compounded with semiconductor photocatalysts, using its regular two-dimensional planar structure as a photocatalyst carrier. On the one hand, this could improve the dispersion of the catalyst. On the other hand, it could accelerate the photogenerated charge migration rate, and improve the photocatalytic activity of the composites.

Huang et al. [64] used the hydrothermal method to add graphene with the mass fractions of 0.1, 0.5, and 1 to Cu_2O, which were noted as Cu_2O/GO-0.1, Cu_2O/GO-0.5, and Cu_2O/GO-1, respectively (Figure 6). The experimental results showed that the highest hydrogen yield of Cu_2O modified with graphene (118.3 mmol) was more than twice that of pure Cu_2O (44.6 mmol). During the formation of Cu_2O/GO composites, many negatively charged functional groups in graphene can recombine with positively charged copper ions by electrostatic adsorption, thus forming Cu_2O/graphene composite structures directly during the reduction process. This principle has been used to synthesize cubic and octahedral Cu_2O/GO composites. This structure could improve the efficiency of electron–hole separation. It could also improve the stability of the prepared catalysts. The experimental results showed that the cubic and octahedral Cu_2O/GO composites degraded methyl orange with more than 90% efficiency. After six replicate tests, the efficiency remained above 70%, indicating that the prepared catalysts had excellent stability [65].

Graphene has properties such as the half-integer Hall effect, a unique quantum tunneling effect, and the bipolar electric field effect. In particular, its excellent electrical conductivity and huge specific surface area provide a feasible way to solve the bottleneck problem in the photocatalytic reaction of Cu_2O-based composites.

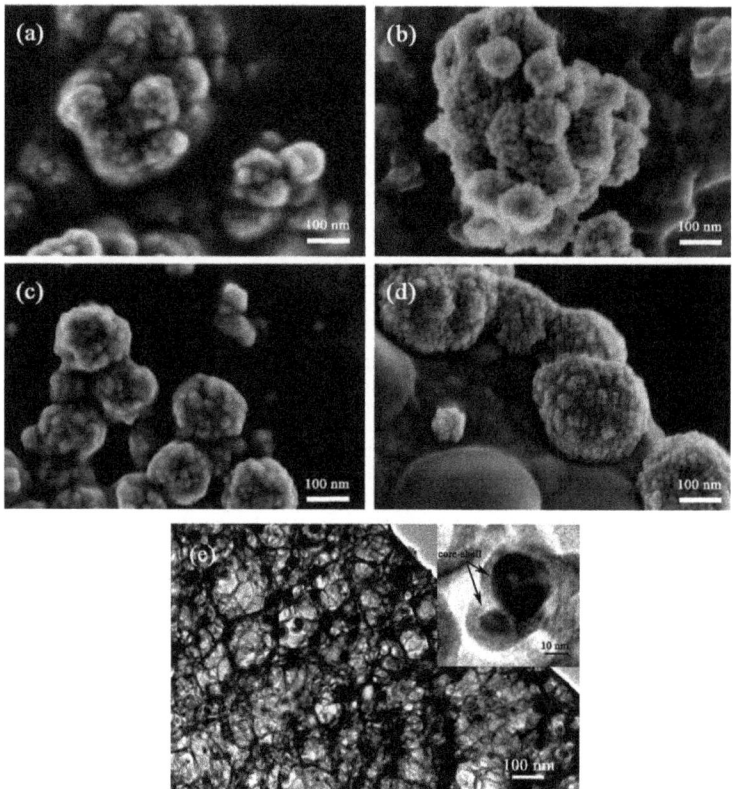

Figure 6. SEM images of (**a**) pure Cu_2O, (**b**) GO/Cu_2O-0.1, (**c**) GO/Cu_2O-0.5, and (**d**) GO/Cu_2O-1, and (**e**) a TEM image of GO/Cu_2O-0.5 [64]. Copyright 2017, Elsevier.

3.2. Ternary Cu_2O-Based Heterojunctions

In recent years, binary photocatalytic composites of Cu_2O have achieved high achievements in the treatment of organic matter form wastewater and CO_2 reduction. However, it will be a long time before binary photocatalytic composites can be used in society and daily life. Therefore, the development of ternary photocatalytic composites has become inevitable.

Yang et al. [66] used ternary Ag-CuO/GO as a photocatalytic material in the photocatalytic degradation of methyl orange, and the degradation efficiency of Ag-CuO/GO on the methyl orange was 90% after 60 min of visible light irradiation. Fu et al. [67] prepared TiO_2-Ag-Cu_2O composite catalysts for enhanced photocatalytic hydrogen production. The experimental results showed that the synergistic effect of Ag and Cu_2O improved the photocatalytic efficiency of the reaction. In addition, the prepared composite catalysts had a double Z-scheme charge transfer pathway, which reduced the electron–hole complexation probability. The weak oxidation holes and weak reduction electrons in the charge transfer process were directly quenched, and the photogenerated carrier separation efficiency and catalyst reduction capacity were significantly enhanced.

3.3. Co-Catalyst Addition

In addition to constructing heterojunction structures, the photocatalytic efficiency can be improved by adding co-catalysts. Suitable co-catalysts are often present on the photocatalyst surface as active centers for oxidation or reduction, which can reduce the oxidation or reduction overpotential, and thus contribute to the photocatalytic reaction. In general, co-catalysts have three primary roles: (1) promoting the separation of the photoexcited electron–hole pairs; (2) inhibiting side reactions; and (3) improving the selectivity of the target products. Yu et al. [68] reported that adding the co-catalyst Cl to Cu_2O nanorods led to a strong CO_2 reduction ability. The experimental results showed that the addition of co-catalyst Cl mainly reduced its direct energy band, and also achieved an increase in the carrier density and conductivity. Zhang et al. [69] doped Zn in Cu_2O microcubes, and the hydrogen production rate of Cu_2O was six times higher than that of pure Cu_2O when the Zn content was 0.1 wt.%. Kalubowila et al. [70] proposed a new method for introducing co-catalysts. They used ascorbic acid (AA) to reduce the prepared Cu_2O/GO, where Cu_2O was partially converted to Cu, and GO was fully converted to rGO. Cu nanoparticles with tens of nanometers have acted as co-catalysts in Cu_2O/Cu/rGO composites, providing centers for effective charge transfer, and enhancing the performance of photocatalytic degradation.

4. Photocatalytic Applications

Semiconductor photocatalytic reactions are based on the solid energy band theory. Under the light, the available photogenerated electrons (e^-) and holes (h^+) in the conduction band (CB) and valence band (VB) of the semiconductor migrate to the surface, to participate in the redox reaction. Therefore, the appropriate match between the CB/VB position of the photocatalyst and the redox potential determines whether the reaction can occur. In general, the CB position of the photocatalyst should be more negative to the reduction potential of the reaction, to promote the transfer of e^- from CB to the reactant; at the same time, the VB position should be corrected to the oxidation potential of the reaction, to ensure that holes can be transported from VB to the reactant. The bandgap of Cu_2O-based materials is shown in Figure 7 [71]. It has been proven that they can be used as photocatalysts to achieve CO_2 reduction (CO_2RR), hydrogen production from water, pollutant degradation, and the reduction reaction of Cr. This section summarizes and discusses the latest progress in applications of Cu_2O-based photocatalysts.

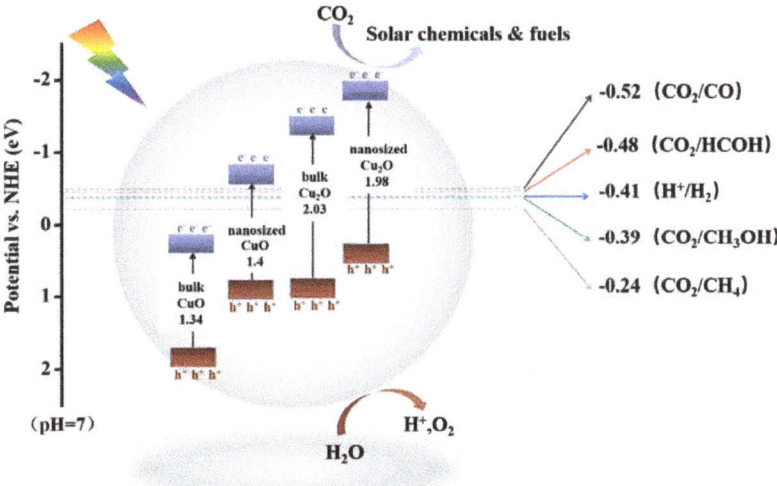

Figure 7. Diagram of the bandgap of copper-oxide-based photocatalysts [71]. Copyright 2022, Elsevier.

4.1. Photocatalytic CO_2 Reduction

Photocatalytic technology can convert CO_2 into CO and hydrocarbon fuels, achieving carbon recycling, and reducing greenhouse gas emissions. The application of Cu_2O has been hampered largely by its inherent photocorrosion, ultra-fast charge recombination rate, and slow charge transport dynamics. In recent years, researchers have conducted and developed a series of novel Cu_2O-based photocatalysts, making significant progress.

As is well known, semiconductors with different morphologies often expose different crystal faces, and exhibit varying photocatalytic activity. Celaya et al. [72] calculated by density-functional theory (DFT) that the (110) and (111) crystal faces of Cu_2O have the potential of photocatalytic reduction of CO_2 to produce hydrocarbon derivatives. To further determine the catalytic mechanism and active site, Wu and his colleagues [73] successfully prepared Cu_2O nanocrystals with (110) and (100) crystal faces through colloidal synthesis, and carried out photocatalytic reactions using CO_2 and H_2O. gas chromatography–mass spectrometry (GC-MS), confirming that methanol was the only product of photoreduction, and the internal quantum yield was approximately 72%. In photocatalytic reactions, the (110) surface of a single Cu_2O particle showed photocatalytic activity, while the (100) surface was inert. The electronic density of the Cu active site on the (110) surface moved from Cu (i) to Cu (ii), and the oxidation state of the Cu changed from Cu (ii) to Cu (i) after CO_2 conversion under light. In 2022, Sahu et al. [74] synthesized and characterized Cu_2O photocatalysts with cubic and truncated cubic structures. Their correspondingly exposed crystal faces were different (Figure 8). Due to the selective accumulation of e^- and h^+ on different crystal planes, the photocatalytic activity in selectively reducing CO_2 to methanol on cubic Cu_2O with anisotropic {100} and {110} crystal planes was nearly 5.5 times higher than that on cubic Cu_2O with only {100} crystal planes.

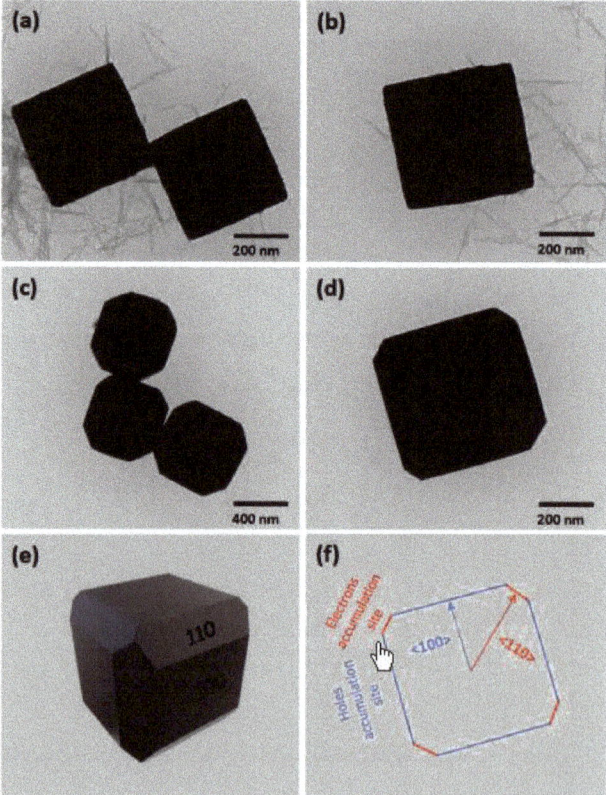

Figure 8. TEM images of (**a**,**b**) cubic Cu_2O and (**c**,**d**) edge-truncated cubic Cu_2O; simulated images of (**e**) 3D structure of edge-truncated cubic Cu_2O and (**f**) 2D crystal orientation [74]. Copyright 2022 American Chemical Society.

Meanwhile, researchers have adopted various modification methods to optimize the structure and performance of the photocatalyst. Element doping is a commonly used method to effectively change the physical properties of semiconductors, to improve their catalytic activity. Cl doping has been shown to optimize the catalytic activity of Cu_2O [75]. At 400 nm, the apparent quantum yield (AQE) of Cl-doped Cu_2O photocatalytic reduction of CO_2 to CO and CH_4 increased, with 1.13% and 1.07% for CO and CH_4, respectively. The reason behind the enhanced performance of CO_2RR was not only that the Cl doping optimized the energy band structure and conductivity of Cu_2O, and improved the adsorption capacity of CO_2 and the separation efficiency of the photogenerated carriers, but also that the Cl-doped Cu_2O was conducive to the conversion of CO_2 into the intermediates of *COOH, *CO, and *CH_3O, thus improving the yield and selectivity of CO and CH_4.

Constructing heterojunction structures is also an effective method for band reconstruction. In heterostructures, the internal electric field is formed at the contact interface of two or more semiconductors with the movement of the Fermi level, which drives the directional migration and separation of photogenerated electrons and holes. Common heterojunctions include the traditional (Type-I, II, and III), p–n, Z-scheme, and S-scheme. The p–n heterojunction of Cu_2O and n-type semiconductors can effectively delay the recombination of photogenerated carriers, and promote electron transfer [76]. The yield of the photocatalytic reduction of CO_2 to CH_3OH from the Cu_2O/TiO_2 heterojunction after 6 h of UV–Vis irradiation has been 21.0–70.6 $\mu mol/g_{cat}$. At the p–n heterojunction, the photogenerated electrons and holes are separated and transferred to the CB/VB with lower potential energy, respectively, resulting in a redox ability closer to the lower of the two semiconductors. The Z-scheme heterostructure solves this problem perfectly. Electrons and holes in the CB/VB with lower energy recombine, and cancel each other out in the Z-scheme heterojunction, thus retaining the higher conduction and valence band values in the two semiconductors, and enhancing the redox ability of the photocatalyst. For example, the Ag-Cu_2O/ZnO nanorods (NRs) reported by Zhang and his team showed an enhanced photocatalytic CO_2 reduction performance [77]. Under UV–vis light, the yield of CO significantly increased, which was seven times higher that of pure ZnO or Cu_2O NRs. The results showed that the deposited Cu_2O can enhance the chemical adsorption of CO_2 on the catalyst surface, and the Z-scheme charge transfer pathway formed between the ZnO and Cu_2O can promote effective charge separation, thereby improving the photocatalysis performance.

Due to the small bandgap energy and high conduction band value of Cu_2O-based materials, the products of photocatalytic CO_2RR are complex, mainly including CO and various organic compounds (CH_4, CH_3OH, HCOOH). According to the different reaction products, the application of Cu_2O-based materials in photocatalytic CO_2 reduction is summarized in Table 1.

Table 1. The application of Cu$_2$O-based photocatalysts for CO$_2$RR.

Photocatalyst	Synthesis Method	Morphology and Structure	Size	Bandgap (E$_g$)	Light Resource	Product	Yield (μmol·g^{-1} h^{-1})	Energy Conversion Efficiency/Selectivity	Refs.
Cu$_2$O/Cu/CVO	Hydrothermal and wet chemical reduction methods	Cu$_2$O nanoclusters and Cu NPs cover the surface of elliptic CVO NPs	~100 nm	Eg$_{CVO}$: 2.34 eV; Eg$_{Cu_2O}$: 1.87 eV	300 W Xe lamp (λ > 400 nm)	CO and CH$_4$	6.97 and 1.62	Selectivity: 51.3% for CO	[78]
3D porous Cu$_2$O	Electrodeposition and thermal oxidation.	3D porous structure	23–25 μm	2.0 eV	300 W Xe lamp (λ > 420 nm)	CO, CH$_4$, and C$_2$H$_4$	26.8, 4.04, and 0.66	—	[79]
Spherical Cu/Cu$_2$O	Solution chemical method	Spherical structure	1 μm	—	300 W Xe lamp (λ > 420 nm)	CO, CH$_3$OH, and H$_2$	87.7, 10.2, and 5.4	—	[80]
Cu$_2$O-Pd	AA reduction and in situ methods	Cube	~2 μm	1.90 eV	300 W Xe lamp (λ > 420 nm)	CO	0.13	—	[81]
Uio-66-NH$_2$/Cu$_2$O/Cu	Hydrothermal method	Octahedron UiO-66-NH$_2$ and Cu attached to the surface of polyhedron Cu$_2$O	1.5 μm	2.79 eV	300 W Xe lamp	CO	4.54	—	[82]
Cu$_2$O-111-Cu^0	One-pot method	Octahedral structure	side length of ~1 μm	1.98 eV	300 W Xe lamp	CH$_4$	78.4	97%	[83]
Ag$_1$/Cu$_2$O@rGO	Water bath combining with gas-bubbling-assisted membrane reduction	Ultrathin rGO nanosheet and Ag NPs supported on Cu$_2$O octahedral nanocrystals	Cu$_2$O: 300 nm; Ag: 10.7 nm; rGO: 1.0 nm (thickness)	Eg$_{Ag/Cu_2O}$: 1.92 eV; Eg$_{Cu_2O}$: 2.0 eV	300 W Xe lamp (λ > 380 nm)	CH$_4$	82.6	AQE: 1.26%. Selectivity: 95.4%	[84]
1D Cu$_2$O@Cu NRs	In situ reduction method	One-dimensional nanorod arrays	<100 nm	2.03 eV	350 W Xe lamp (λ > 420 nm)	CH$_4$ and C$_2$H$_4$	—	AQE: 2.4%	[85]
RT-Cu$_{0.75}$	Low temperature thermochemical reduction and photo-deposition	—	—	2.72 eV	100 W solar simulator with an AM 1.5 filter	CH$_4$	77 nmol·g^{-1} h^{-1}	AQE: 0.012%	[86]
U-Cu$_2$O-LTH@PCN-X	In situ reduction	Ultrafine nanoclusters	<3 nm	Eg$_{PCN}$: 2.62 eV; Eg$_{U-Cu_2O-LTH}$: 2.07 eV	300 W Xe lamp (λ > 400 nm)	CH$_3$OH	51.22	AQE: 1.01%	[87]

Table 1. *Cont.*

Photocatalyst	Synthesis Method	Morphology and Structure	Size	Bandgap (E_g)	Light Resource	Product	Yield ($\mu mol \cdot g^{-1} \cdot h^{-1}$)	Energy Conversion Efficiency/Selectivity	Refs.
Fe_3O_4@N-C/Cu_2O	AA reduction and aerobic oxidation	Rod-shaped core-shell nanostructure	5 nm (thickness of NC shell layer)	—	5 W Xe HID lamp	CH_3OH	146.7	—	[88]
Dodeca-Cu_2O/rGO	Solution-chemistry	Rhombic dodecahedra	400–700 nm	2.16 eV	300 W Xe lamp (λ > 420 nm)	CH_3OH	17.765	—	[89]
Carbon layer@CQDs/Cu_2O	Hydrothermal method	Nearly spherical structure	~2 μm diameter	2.09 eV	300 W Xe lamp	CH_3OH	99.6	—	[90]
Ti_3C_2 QDs/Cu_2O NWs/Cu	Self-assembly strategy	QDs incorporated onto NWs	~500 nm (diameter of NWs)	2.02 eV	AM 1.5, 300 W Xe lamp	CH_3OH	78.50	—	[91]
Cu@Cu_2O	Thermal treatment	Core-shell nanoparticles	~70 nm diameter	—	Xe lamp (420–780 nm)	HCOOH	67.35	AQE: 0.12% at 560 nm	[92]
NH_2-C@Cu_2O	Low temperature annealing	Octahedral structure	—	1.79 eV	300 W Xe lamp (λ > 420 nm)	HCOOH	138.65	Selectivity: 92%	[93]

4.2. Photocatalytic H_2 Production

Hydrogen energy is abundant and renewable, which can effectively avoid energy exhaustion, and the products of hydrogen energy combustion will not cause pollution. Photocatalytic hydrogen production has the advantages of high efficiency, low cost, and environmental friendliness, and has great potential in high-efficiency hydrogen evolution. Common semiconductor photocatalysts (such as TiO_2, ZnO, and g-C_3N_4.) have the disadvantage of a low utilization of sunlight, and the photocatalytic hydrogen evolution efficiency is not ideal. In recent years, Cu_2O has become a research hotspot in photocatalytic hydrogen evolution because of its excellent photoresponsiveness. However, the poor charge separation ability of pure Cu_2O lowers its hydrogen evolution performance. It is essential to modify and adjust Cu_2O-based catalysts to meet the practical need to increase the hydrogen production yield.

Hybridizing Cu_2O with other semiconductor materials to construct heterojunctions can achieve the effective separation of photo-induced charge carriers, which is an effective method to enhance photocatalytic activity, and has been validated in numerous studies on photocatalytic hydrogen production. $NiFe_2O_4$/Cu_2O with different mass percentages has been synthesized by impregnation and thermal annealing methods to construct p–n heterojunctions [94]. The photocatalytic hydrogen production rate of all heterojunctions was significantly higher than that of the original material. The 50/50 mass ratio was the most effective, and the hydrogen production rate within 24 h was 102.4 mmol·g^{-1}, while $NiFe_2O_4$ and Cu_2O only obtained 1.35 and 0.85 mmol·g^{-1}, respectively. The increase in activity came from the enhanced charge separation at the heterojunction, which increased the concentration of charge carriers (Figure 9). Cu_2O/$CaTiO_3$ series samples were synthesized using the hydrothermal method and $NaBH_4$ reduction treatment [95]. The photocatalytic hydrogen production effect of the 50Ca10Cu sample was the best (8.268 mmol·g^{-1}·h^{-1}), about 344.5 times that of the $CaTiO_3$ sample. It also exhibited perfect stability after multiple cyclic tests.

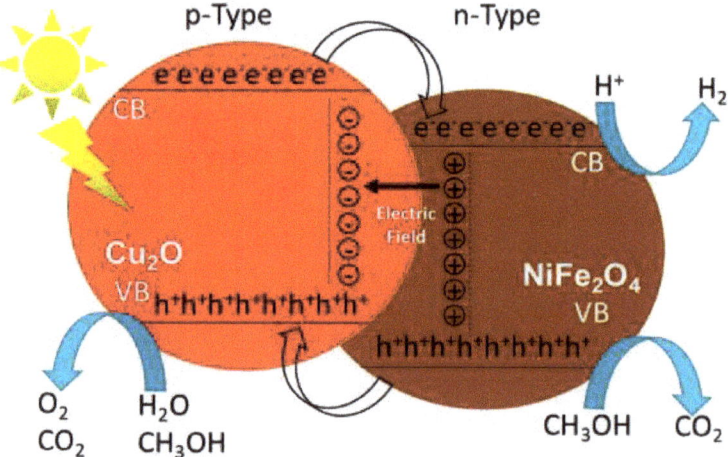

Figure 9. The p–n heterostructures in the $NiFe_2O_4$/Cu_2O photocatalyst [94].

The above p–n heterojunctions are typical type-II heterojunctions, which often impair the redox capacity of photogenerated electrons and holes. Researchers have recently designed and constructed Z-scheme and step-scheme (S-scheme) heterojunctions for photocatalytic hydrogen production. For example, dendritic branched Cu_2O was synthesized hydrothermally, and Cu_2O/TiO_2 composites were prepared via surface charge modulation [96]. The hydrogen production rate of the optimized CT-70 (Cu_2O coupled with 70 wt.% TiO_2) photocatalyst reached 14.020 $mmol^{-1}$ within six hours, which was 264 and

44 times higher than that of pure Cu_2O and TiO_2, respectively. The electron transfer mechanism of the Z-scheme was proposed and verified via DFT calculation and EPR analysis. Under simulated sunlight, photoexcited electrons migrate from the CB of TiO_2 to the VB of Cu_2O, and then recombine with photogenerated holes in the VB of Cu_2O, thereby retaining highly reducing electrons and highly oxidizing holes (Figure 10). Therefore, under the conditions of sensitive photosensitivity and the effective separation of photogenerated electrons and holes, the performance of photocatalysts in hydrogen evolution under visible light is significantly improved. The S-scheme heterojunction photocatalyst has a similar efficient carrier separation performance and enhanced redox capacity. $Cu_2O/g-C_3N_4$ composites were successfully synthesized using a simple wet chemical method, and applied in the field of photocatalytic energy production. $Cu_2O/g-C_3N_4$ series samples showed high catalytic activity. In particular, 1-$Cu_2O/g-C_3N_4$ showed the highest hydrogen evolution rate of 480.6 $\mu mol \cdot g^{-1} \cdot h^{-1}$ under visible light irradiation, 12.0 times that of the original Cu_2O sample. Based on the analysis of the experimental and simulation results, the ideal catalytic performance of the $Cu_2O/g-C_3N_4$ photocatalyst was derived from the efficient interfacial charge separation and transfer of the S-scheme heterostructure [97].

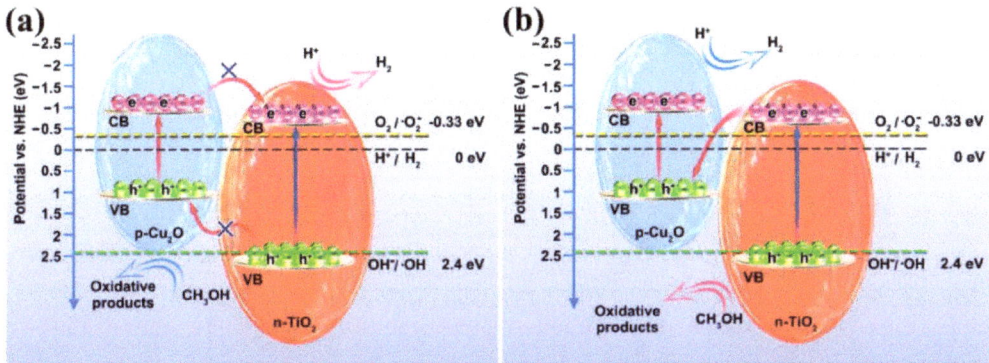

Figure 10. (a) Type-II and (b) Z-scheme electron transfer mechanism in Cu_2O/TiO_2 photocatalyst [96]. Copyright 2021, Elsevier.

Furthermore, photocorrosion is currently an urgent problem for Cu_2O photocatalysts, and finding effective strategies to suppress photocorrosion in photocatalysts is still an enormous challenge. To overcome this challenge, Liu et al. [98] proposed a core–shell model: the Cu_2O/PyTTA-TPA COF nanocube photocatalyst was constructed using an energy level matching the Cu_2O and 2D PyTTA-TPA COF. It exhibited an excellent photocatalytic hydrogen evolution rate of 12.5 $mmol \cdot g^{-1} \cdot h^{-1}$, approximately 8.0 and 20.0 times higher than the PyTTA TPA COF and Cu_2O, respectively. Most importantly, under the protection of the stable PyTTATPA-COF shell, the Cu_2O nanocube core was protected from photocorrosion, and did not show noticeable morphological or crystal structure changes after 1000 light excitations, thus significantly improving the photocorrosion resistance stability of the catalyst. Table 2 shows the recently reported Cu_2O-based materials for photocatalytic hydrogen production.

Table 2. The reported applications of Cu_2O-based materials in photocatalytic hydrogen production in recent years.

Photocatalyst	Synthesis Method	Morphology and Structure	Size	Bandgap (E_g)	Light Resource	Yield ($\mu mol \cdot g^{-1} \cdot h^{-1}$)	Energy Conversion Efficiency/Selectivity	Refs.
Cu_2O/TiO_2	Ball-milling	Irregular shapes	Anatase: 16.2 nm Rutile: 30.5 nm	3.08 eV	High-pressure Hg lamp (125 W)	200	AQE: 1.51% Light-to-chemical energy efficiency: 0.6%	[99]
$In(OH)_3$-In_2S_3-Cu_2O	Hydrothermal, wet chemical and electrospinning process	Nanofiber	100–200 nm of diameter	$Eg_{In(OH)_3}$: 5.15 eV $Eg_{In_2S_3}$: 1.98 eV Eg_{Cu_2O}: 2.17 eV	5 W blue light LED (λ_{max} = 420 nm)	1786.5	—	[100]
$Au@Cu_2O$	Sequential ion-exchange reaction	Core-shell architectures	54.4 ± 4.8 nm	Eg_{Cu_2O}: 2.40 eV	Xenon lamp and AM 1.5G filter	55.5	AQE: 0.29% at 420 nm	[101]
$Na_2Ti_6O_{13}/CuO/Cu_2O$	Solid-state and impregnation method	Belt morphology	1 μm	3.61 eV	UV/vis lamp (254 nm, 4400 μW/cm^2)	33	—	[102]
$C@Cu_2O/CuO$	Calcination	Chrysanthemum-like crystalline	—	2.0 eV	350 W Xe lamp (40 mW/cm^2)	26,700	External quantum efficiency (EQE): 52.4%	[103]
$NiCo$-LDH/Cu_2O	Electrostatic self-assembly	3D flower cluster	—	$Eg_{NiCo-LDH}$: 1.78 eV Eg_{Cu_2O}: 1.89 eV	5 W LED ($\lambda \geq$ 420 nm)	3666	—	[104]
Cu_2O/TiO_2	DES-assisted synthesis	Cu_2O nanoclusters on TiO_2 surfaces	1.5 nm of Cu_2O nanoclusters and 25.8 nm of TiO_2 particles	Eg_{TiO_2}: 3.12 eV Eg_{Cu_2O}: 2.13 eV	300 W Xe lamp	24,210	—	[105]
$Cu@TiO_2$-Cu_2O	Hydrothermal and $NaBH_4$ treatment	Urchin-like hierarchical spheres	—	Eg_{TiO_2}: 3.18 eV Eg_{Cu_2O}: 2.05 eV	300 W Xenon lamp	12,000.6	AQE: 8.26%	[106]
Cu/Cu_2O	Microwave-assisted heating	Hollow spherical morphology	430 ± 1.2 nm in diameter	2.0 eV	LED light (20 W)	141	—	[107]
$Cu_2O/SiO_2/CdIF$	Reactive deposition	Core-shell structure	—	Eg_{CdIF}: 5.09 eV Eg_{Cu_2O}: 2.22 eV	300 W xenon lamp (340–780 nm)	2879.09	AQE: 0.040% at 420 nm	[108]

4.3. Photocatalytic Degradation of Pollutants

With the rapid development of the global economy, industrial and agricultural waste is produced in large quantities, and continues to enter the environment. Many organic pollutants also enter the environment, and some show persistent pollution, which is difficult to remove through microbial action and hydrolysis. The long-term existence and accumulation of refractory pollutants leads to environmental pollution and ecological imbalance, and even threatens human survival and development. Research and development around pollutant degradation technology are critical. Photocatalytic technology has shown promising prospects for treating refractory pollutants, such as the photocatalytic processes that mineralize organic pollutants into water and CO_2, and which essentially eliminate secondary pollution, rather than concentrating these pollutants and their by-products into the waste stream. In the past few decades, extensive research has been conducted on Cu_2O-based photocatalysts to purify the environment. Table 3 summarizes the recent reports of Cu_2O-based photocatalysts in pollutant degradation.

Among all the types of pollutants, organic dyes have become an important source of water pollution. As refractory organic pollutants, dyes cause severe damage to human health and the ecological balance. Traditional adsorption methods only transfer toxic organic molecules to the solid surface, without eliminating them, and still run the risk of desorption. MBC@Cu_2O composites have been prepared by loading porous spherical Cu_2O onto wood biochar carriers, with a liquid-phase synthesis strategy, at room temperature [109]. As a bi-functional adsorption-based photocatalytic composite, MBC@Cu_2O showed great potential in removing anionic dye methyl orange (MO) from water. Under visible light irradiation, the photocatalytic degradation efficiency of MO reached 94.5%, and remained above 80% after five cycles. In another work, Sehrawat and his team prepared MoS_2/Cu_2O composites with different weight ratios via precipitation, using MoS_2 nanosheets and Cu_2O nanospheres [110]. The photocatalytic degradation of indigo carmine (IC) dye was carried out under simulated visible light. Compared to the original MoS_2, the optimized MC-3 sample showed the best degradation performance, with a degradation rate of 99.59% for IC within 90 min, and no significant change in performance after five cycles. Experiments regarding the capture of active species showed that the photocatalytic reaction relied on the production of the superoxide radical ($\bullet O_2^-$), and further verified the Z-scheme mechanism of the MoS_2/Cu_2O photocatalyst. In the same year, Li et al. synthesized the core–shell WO_3-Cu_2O Z-scheme heterojunction via hydrothermal and electrochemical deposition methods for the photocatalytic degradation of methylene blue (MB) under visible light [111]. The Cu_2O nanoparticles deposited on the surface of WO_3 enhanced the visible light absorption ability. The Z-scheme heterojunction achieved the effective spatial separation of the charges, and retained the strong redox ability of the photogenerated electrons and holes. The WO_3-Cu_2O-120s photocatalyst showed the highest reaction rate, almost twice that of the original WO_3.

As a typical persistent organic pollutant, antibiotics are difficult to degrade and remove, due to their low biodegradability, which has become a thorny problem in water pollution control. Research has shown that the defect states and vacancies caused by element doping significantly impact the catalytic performance of semiconductor materials. Doping semiconductor functional materials with specific elements provides a feasible way to overcome the obstacles in applications for photocatalytic degradation. Nie et al. synthesized Cl-doped Cu_2O microcrystals using a simple hydrothermal method, and used them to treat levofloxacin contaminants (LVX) under mild reaction conditions [112]. Compared with other reaction systems, the synthesis of Cl-doped Cu_2O has a higher degradation efficiency for levofloxacin. After 240 min of photocatalytic reaction, the maximum degradation rate of LVX was 85.8% and 80.3% after eight cycles, indicating the stability and reusability of the photocatalyst. Based on the theoretical calculation and test results, it can be concluded that introducing hybrid orbitals and oxygen vacancy defects into Cu_2O crystal cells by doping Cl reduces the band gap of Cu_2O, resulting in a red shift in the absorption edge. Compared with pure Cu_2O microcrystals, the prepared Cl-doped Cu_2O single crystals with oxygen vacancy had a narrower band gap, and higher photogenerated electron–hole separation and transport efficiency. Considering the close relationship between the morphology and electronic structure, surface energy, and chemical reactivity of nanocrystals, it is of great significance to explore the influence of the morphology/exposed crystal surface of Cu_2O on the synthesis process and the photocatalytic performance. Wu et al. developed a series of Cu_2O@HKUST-1 core–shell structures via self-constrained strategies, using Cu_2O nanocrystals with different morphologies as templates [113]. The characterization results indicated that the (111) surface of Cu_2O was more favorable for the growth of HKUST-1 than the (100) surface. Comparing the photocatalytic degradation performance of tetracycline hydrochloride (TC-HCl), it was found that Cu_2O@HKUST-1 had the best photocatalytic performance among the three types of composite material, with a degradation efficiency of 95.35% for TC-HCl. It was attributed to the excellent photoresponse, and the most effective interfacial charge transfer and separation in the Cu_2O@HKUST-1 cubes.

In addition to organic dyes and antibiotics, solar-powered Cu_2O-based photocatalysts can degrade heavy metal pollutants in wastewater, mainly toxic hexavalent chromium (Cr (VI)). Xiong et al. [114] constructed a Cu_2O/LDH photocatalyst by grafting Cu_2O-NP, and embedding it into the LDH host layer through an in-situ reduction strategy. CuZnTi LDH is valuable in two aspects: (a) as a source of Cu_2O, and (b) as a support bracket to avoid the self-oxidation of Cu_2O-NPs. The optimized photocatalyst showed a high degradation efficiency for difficult-to-degrade pollutants under visible light conditions, with a reduction rate of 95.5% for Cr (VI) by Cu_2O/LDH0.10, and a degradation rate of 71.6% for TC. The excellent photocatalytic efficiency was attributed to the charge transfer mechanism of the Cu_2O/ZnTiLDH p–n heterojunction, effectively promoting the separation and migration of the photogenerated electron–hole. Recently, Zhu et al. [115] used the Si and Cu of waste serpentine tailings and WPCB to prepare low-cost waste-based Cu-Cu_2O/SiO_2 photocatalysts. Due to the dispersion of Cu-Cu_2O_3 on the surface of the SiO_2 carrier, the composite material obtained a higher specific surface area. The photocatalytic reduction of Cr (VI) using waste-based catalysts was the best at a loading rate of 9% Cu and $7g·L^{-1}$ SiO_2, and the photocatalytic activity decreased by only 4.93% after five cycles. The mechanism of Cr (VI) reduction by the waste Cu-Cu_2O/SiO_2 photocatalyst is to excite the waste Cu_2O to produce photoelectron–hole pairs. The electrons in the waste group Cu_2O CB reduce Cr (VI) adsorbed on the surface to Cr (III), and the surface Cu drives the electrons to the surface of the Cu metal, without returning the waste group Cu_2O.

Moreover, the accumulation in soil and water of herbicides, insecticides used in the agriculture and food industries, and phenolic compounds emitted from industry, such as petrochemicals and pharmaceuticals, can have significant harmful effects on humans and aquaculture systems. The use of metal oxide photocatalysts has been proven to be an effective, low-cost, and green method for treating such wastewater. In 2021, Alp [116] successfully synthesized hybrid Cu_2O-Cu cubes by reducing D(+)-glucose in an alkaline solution using a one-step aqueous solution synthesis method, without any toxic reagents or surfactants. The Cu_2O-Cu exhibited excellent photocatalytic properties for dyes and herbicides, due to the effective separation of photogenerated electron–holes and the enhanced charge transfer mechanism at heterojunctions. In particular, when dealing with 2,4-Dichlorophenoxyacetic acid (2,4-D), one of the widely used herbicides in agriculture and urban landscaping, the degradation effect of the Cu_2O-Cu heterojunction was outstanding. It photodegraded all of the 2,4-D in the medium within 40 min, while the original Cu_2O cube photodegraded 85% within 60 min. In the same year, Mkhalid et al. [117] prepared a Cu_2O photocatalyst loaded with Cu nanoparticles via sol–gel and photo-assisted deposition technology. The structure and optical and photoelectric properties of the prepared photocatalyst were improved by adjusting the Cu content. The results showed that the band gap of the Cu_2O loaded with 15% Cu was reduced to 1.95 eV, significantly enhancing the visible light absorption ability. The optimized Cu@Cu_2O photocatalyst completely photodecomposed atrazine (AZ, a commonly used triazine herbicide) within 30 min, and demonstrated excellent durability. In recent years, effectively solving the problem of phenolic pollutants in livable environments has also been a major challenge faced by humanity, and has received a high level of attention from many researchers. A low-cost but highly efficient phosphate-doped carbon/Cu$_2$O composite (HKUST-1-P-300) was reported by Dubai et al. [118]. The catalyst was derived from the modification of HKUST-1 with triphenylphosphine and conditioned calcination. Under visible light irradiation, the degradation efficiency of HKUST-1-P-300 for phenol was 99.8%, the hydrogen evolution rate was 1208 μmol, and the external quantum efficiency was 48.6% (at 425 nm) within 90 min, and the high performance could still be maintained after four cycles. Mechanism studies showed that the excellent photocatalytic activity of HKUST-1-P-300 came from multiple synergistic effects: an enhanced visible light absorption efficiency, a larger surface area, the effective separation of photogenerated carriers, a reduced aggregation of Cu$_2$O, and the P-doped carbon/Cu$_2$O structure. These novel Cu$_2$O-based materials, as highly efficient photocatalysts, have potential applications in removing environmental pollutants, and generating clean energy, to promote sustainable environmental construction.

Table 3. Recently reported Cu$_2$O-based materials for the photocatalytic removal of pollutants.

Photocatalyst	Synthesis Method	Morphology and Structure	Size	Bandgap (Eg)	Light Resource	Target Pollutant/Concentration/Volume	Efficiency	Cycle	Refs.
Ag-Cu$_2$O	Electrochemical deposition and redox reaction	Composite film	—	2.02 eV	500 W halogen lamp	MB/30 mg·L^{-1}/50 mL	92%	3	[119]
Cr-doped Cu$_2$O	Hydrothermal method	Octahedrons	800–1200 nm	2.06 eV	500 W tungsten halogen lamp (400–1100 nm)	LVX/40 mg·L^{-1}/50 mL	79.6–72.4%	1–8	[120]
BiOCl/Cu$_2$O	Solvothermal method	Spherical shape	3–5 μm	2.00 eV	500 W Xenon lamp	Moxifloxacin/20 mg·L^{-1}/50 mL	72.3%	5	[121]
C-dots/Cu$_2$O/SrTiO$_3$	Hydrothermal and two-step method	Chocolate ball with sesame on the surface	~2.16 μm	Eg$_{SrTiO_3}$: 3.19 eV; Eg$_{Cu_2O}$: 2.10 eV	500 W Xenon lamp (λ > 420 nm)	CTC·HCl/15 mg·L^{-1}/50 mL	92.6%	4	[122]
CuO-Cu$_2$O	Chemical–thermal oxidation	Nanorods	60 nm	1.90 eV	150 W metal halide lamp (λ > 400 nm)	MB/5 mg·L^{-1}/50 mL	80%	3	[123]
Cotton fabrics/Cu$_2$O-NC	Impregnation and HH reduction	Octahedron Cu$_2$O attached to cotton fibers	20–40 nm of diameter of Cu$_2$O	Eg$_{Cu_2O}$: 2.20 eV	350 W Xenon lamp (λ > 400 nm)	MB/200 ppm/200 mL	98.32–85%	1–5	[124]
Cu$_2$O@HKUST-1	In-situ converted strategy	Octahedron structure	—	Eg$_{Cu_2O}$: 1.95 eV; Eg$_{HKUST-1}$: 2.59 eV	Tungsten lamp (>420 nm, 500 W)	TC-HCl/20 mg·L^{-1}/100 mL	93.40–90.02%	1–4	[125]
Fe$_3$O$_4$/Cu$_2$O-Ag	Solvothermal and liquid deposition methods	Double six peak structure	~5 nm	2.23 eV	—	PAHs/5 mg·L^{-1}/100 mL	95–90%	1–8	[126]
Cu$_2$O/ZnO@PET	Electroless template deposition	Rectangular-shaped	~13 ± 4.5 nm	3.2–3.4 eV	Ultra-Vitalux 300W	Czm/1.0 mg·L^{-1}/100 mL	98–26%	1–6	[127]
Cu$_2$O-Au-TiO$_2$	Two-step photocatalytic deposition	Core-shell structure	~50 nm	1.4–1.7 eV	Xenon lamp (λ > 422 nm)	Cr(VI)/10 mg·L^{-1}/50 mL	100% (3h)	3	[128]
Cu$_2$O/N-CQD/ZIF-8	Reduction precipitation	Spherical structure	~80–100 nm	2.6 ␣eV,	300 W Xenon lamp (λ > 420 nm)	Cr(VI)/20 mg·L^{-1}/50 mL	98.99–97.13%	1–5	[129]
Cu$_2$O/rGO/BiOBr	Two-step strategy	Hierarchical microspheres	500 nm–1 μm	Eg$_{BiOBr}$: 2.7 eV; Eg$_{Cu_2O}$: 1.9 eV	300 W Xenon lamp (λ > 420 nm)	Cr(VI)/20 mg·L^{-1}/50 mL	100% (40 min)	5	[130]
Cu-TiO$_2$-Cu$_2$O	Photodeposition	The triple junction structure	~20 nm	—	300 W Xenon lamp (200–2400 nm)	2,4,5-T/50 ppm/100 mL	93%	3	[131]
Ag-Cu$_2$O/rGO	Two-step reduction process	Spherical AgNPs deposited on the Cu$_2$O situated on the surface of rGO sheets	~60 nm	—	60 W tungsten filament lamp (500–700 nm, 0.24 W/cm^2)	MO/40 mg·L^{-1}/50 mL Phenol/20 mg·L^{-1}/50 mL	90% (60min); Rate constant of phenol degradation: 0.09732	3	[132]

4.4. DFT Study Applied in the Photocatalysis

At the present time, there are fewer studies revealing the reaction mechanism of Cu$_2$O through DFT simulations. Moreover, the catalytic microstructure and mechanism of Cu$_2$O-based composites are still unclear. Designing Cu$_2$O-based photocatalysts, and investigating the mechanism of improving photocatalytic activity at the molecular level require the introduction of theoretical calculations. In future studies, DFT simulations and experiments are needed, to reveal the relationship between the establishment of the microstructure and the catalytic activity of the photocatalysts, which will provide the theoretical basis for future photocatalytic industrial applications.

Lv et al. [133] analyzed the electronic structure and photocatalytic properties of Cu_2O doped with different contents of Mn, using first-principle calculations. The simulation results showed that the visible light absorption intensity and photocatalytic efficiency were enhanced with the increase in doping concentration, and varied with the doping configuration, compared to pure Cu_2O. The enhanced light absorption was mainly attributed to the in-band leaps of the electrons in the three-dimensional state of Mn. The enhancement of light absorption was mainly due to the in-band leaps of electrons in the three-dimensional state of Mn, which gave the semiconductor material certain metallic properties, and increased the absorbance of the visible light. Therefore, Cu_2O applied to the future industrialization of photocatalysis could be doped with a small amount of Mn in the semiconductor, to improve the photocatalytic efficiency.

5. Conclusions

In recent years, the practical photocatalytic applications of Cu_2O-based materials in scientific fields such as solar energy conversion and environmental remediation have attracted great interest. As a transition metal oxide, Cu_2O has the advantages of a narrow band gap, strong visible light response, suitable conduction band position, low cost, and great potential as a photocatalyst. This paper introduces the basic properties, synthesis methods, and modification strategies of Cu_2O-based materials. Recently reported Cu_2O-based photocatalysts and their recent advances in photocatalysis, such as photocatalytic CO_2 reduction, photocatalytic hydrogen production, and pollutant degradation, are reviewed. However, the research on Cu_2O-based materials is still in its early stages, and there is room for improvement in their photocatalytic performance.

1. Currently, most Cu_2O-based composites and sacrificial agents are synthesized from noble metal materials, which have high costs and significantly limit their large-scale applications. The development of non-precious metal catalysts, such as graphene, is vital to future development. More importantly, the catalytic efficiency of most Cu_2O-based composites is very low, and the catalytic performance needs to be improved to meet the requirements of practical applications.
2. Although many experimental studies on the photocatalysis of Cu_2O-based composites are introduced in this paper, these works are still in their infancy. In addition, the large-scale production of high-quality Cu_2O-based photocatalysts faces numerous difficulties, considering the secondary hazards of nanomaterials. Therefore, it is urgent that we further study the photocatalytic mechanism of Cu_2O-based composites from the above perspectives, and promote the industrial application process of Cu_2O-based composite catalysts.
3. The photocorrosion of Cu_2O still deserves attention. Although the current method of constructing heterojunctions to suppress photocorrosion has achieved certain results, the photocorrosion phenomenon of Cu_2O still exists, and affects its long-term use. Establishing a core–shell structure is a good governance measure but, when synthesizing photocatalysts, it is necessary to carefully handle the thickness of the shell layer, to ensure sufficient absorption of light by the Cu_2O.
4. The structure of the catalyst determines the catalytic activity, while the catalytic microstructure and mechanism of Cu_2O-based composites is still unclear. Theoretical calculations should be introduced when designing a Cu_2O-based photocatalyst, and studying the mechanism of improving photocatalytic activity at the molecular level. In future research, DFT simulations and experiments are needed, to reveal the relationship between the establishment of the microstructure and the catalytic activity of photocatalysts.

Author Contributions: Q.S.: conceptualization, methodology, software, investigation, writing—original draft. C.Z.: methodology, validation, formal analysis, and visualization. M.L. and X.T.: funding, acquisition, and supervision. All authors have read and agreed to the published version of the manuscript.

Funding: Financial support in carrying out this work was provided by the Doctoral Research Foundation of Weifang University (2022BS13).

Institutional Review Board Statement: Not applicable.

Informed Consent Statement: Not applicable.

Data Availability Statement: Not applicable.

Conflicts of Interest: The authors declare no conflict of interest.

Sample Availability: Samples of the compounds are available from the authors.

References

1. Ju, L.; Tang, X.; Zhang, Y.; Li, X.; Cui, X.; Yang, G. Single Selenium Atomic Vacancy Enabled Efficient Visible-Light-Response Photocatalytic NO Reduction to NH_3 on Janus WSSe Monolayer. *Molecules* **2023**, *28*, 2959. [CrossRef]
2. Duic, N.; Guzovic, Z.; Vyatcheslav, K.; Klemes, J.J.; Mathiessen, B.V.; Yan, J.Y. Sustainable development of energy, water and environment systems. *Appl. Energ.* **2013**, *101*, 3–5. [CrossRef]
3. Ju, L.; Tang, X.; Li, J.; Dong, H.; Yang, S.; Gao, Y.; Liu, W. Armchair Janus WSSe Nanotube Designed with Selenium Vacancy as a Promising Photocatalyst for CO_2 Reduction. *Molecules* **2023**, *28*, 4602. [CrossRef] [PubMed]
4. Yang, X.G.; Wang, D.W. Photocatalysis: From fundamental principles to materials and applications. *ACS Appl. Energy Mater.* **2018**, *1*, 6657–6693. [CrossRef]
5. Albero, J.; Peng, Y.; Garcia, H. Photocatalytic CO_2 reduction to C2+ products. *ACS Catal.* **2020**, *10*, 5734–5749. [CrossRef]
6. Fujishima, A.; Honda, K. Electrochemical photolysis of water at a semiconductor electrode. *Nature* **1972**, *238*, 37–38. [CrossRef]
7. Henderson, M.A. A surface science perspective on TiO_2 photocatalysis. *Surf. Sci. Ref.* **2011**, *66*, 185–297. [CrossRef]
8. Nakata, K.; Fujishima, A. TiO_2 photocatalysis: Design and applications. *J. Photoch. Photobio. C* **2012**, *13*, 169–189. [CrossRef]
9. Ma, D.G.; Zhai, S.; Wang, Y.; Liu, A.A.; Chen, C.C. TiO_2 photocatalysis for transfer hydrogenation. *Molecules* **2019**, *24*, 330. [CrossRef]
10. Guo, Q.; Zhou, C.Y.; Ma, Z.B.; Yang, X.M. Fundamentals of TiO_2 photocatalysis: Concepts, mechanisms, and challenges. *Adv. Mater.* **2019**, *31*, 1901997. [CrossRef] [PubMed]
11. Goktas, S.; Goktas, A. A comparative study on recent progress in efficient ZnO based nanocomposite and heterojunction photocatalysts: A review. *J. Alloy Compd.* **2021**, *863*, 158734. [CrossRef]
12. Liu, X.Q.; Iocozzia, J.; Wang, Y.; Cui, X.; Chen, Y.H.; Zhao, S.Q.; Li, Z.; Lin, Z.Q. Noble metal-metal oxide nanohybrids with tailored nanostructures for efficient solar energy conversion, photocatalysis and environmental remediation. *Energy Environ. Sci.* **2017**, *10*, 402–434. [CrossRef]
13. Liu, Y.P.; Shen, S.J.; Zhang, J.T.; Zhong, W.W.; Huang, X.H. $Cu_{2-x}Se$/CdS composite photocatalyst with enhanced visible light photocatalysis activity. *Appl. Surf. Sci.* **2019**, *478*, 762–769. [CrossRef]
14. Sun, Y.B.; Xiao, J.T.; Huang, X.S.; Mei, P.; Wang, H.H. Boosting photocatalytic efficiency of MoS_2/CdS by modulating morphology. *Environ. Sci. Pollut. Res.* **2022**, *29*, 73282–73291. [CrossRef] [PubMed]
15. Ma, M.X.; Jin, H.M.; Wu, Z.K.; Guo, Y.N.; Shang, Q.K. Selective photocatalytic oxidation of aromatic alcohols using B-g-C_3N_4/Bi_2WO_6 composites. *Sep. Purif. Technol.* **2023**, *317*, 123915. [CrossRef]
16. Qin, N.B.; Zhang, S.F.; He, J.Y.; Long, F.; Wang, L.L. In situ synthesis of $BiVO_4$/BiOBr microsphere heterojunction with enhanced photocatalytic performance. *J. Alloy Compd.* **2022**, *927*, 166661. [CrossRef]
17. Jiang, J.Z.; Xiong, Z.G.; Wang, H.T.; Liao, G.D.; Bai, S.S.; Zou, J.; Wu, P.X.; Zhang, P.; Li, X. Sulfur-doped g-C_3N_4/g-C_3N_4 isotype step-scheme heterojunction for photocatalytic H_2 evolution. *J. Mater. Sci. Technol.* **2022**, *118*, 15–24. [CrossRef]
18. Sekkat, A.; Bellet, D.; Chichignoud, G.; Munoz-Rojas, D.; Kaminski-Cachopo, A. Unveiling key limitations of ZnO/Cu_2O all-oxide solar cells through numerical simulations. *ACS Appl. Energy Mater.* **2022**, *5*, 5423–5433. [CrossRef]
19. Wu, L.K.; Ma, P.D.; Zhang, C.H.; Yi, X.K.; Hao, Q.L.; Dou, B.J.; Bin, F. Effects of Cu_2O morphology on the performance of CO self-sustained catalytic combustion. *Appl. Catal. A-Gen.* **2023**, *652*, 119034. [CrossRef]
20. Li, J.W.; Sun, Z.L.; He, M.Z.; Gao, D.; Li, Y.T.; Ma, J.J. Simple synthesis of Ag nanoparticles/Cu_2O cube photocatalyst at room temperature: Efficient electron transfer improves photocatalytic performance. *Inorg. Chem. Commun.* **2022**, *138*, 109200. [CrossRef]
21. Yang, X.; Cheng, J.; Yang, X.; Xu, Y.; Sun, W.F.; Zhou, J.H. MOF-derived Cu@Cu_2O heterogeneous electrocatalyst with moderate intermediates adsorption for highly selective reduction of CO_2 to methanol. *Chem. Eng. J.* **2022**, *431*, 134171. [CrossRef]
22. Wang, N.; Tao, W.; Gong, X.Q.; Zhao, L.P.; Wang, T.S.; Zhao, L.J.; Liu, F.M.; Liu, X.M.; Sun, P.; Lu, G.Y. Highly sensitive and selective NO_2 gas sensor fabricated from Cu_2O-CuO microflowers. *Sens. Actuators B-Chem.* **2022**, *362*, 131803. [CrossRef]
23. Li, J.W.; He, M.Z.; Yan, J.K.; Liu, J.H.; Zhang, J.X.; Ma, J.J. Room temperature engineering crystal facet of Cu_2O for photocatalytic degradation of methyl orange. *Nanomaterials* **2022**, *12*, 1697. [CrossRef] [PubMed]
24. Zhang, Y.H.; Liu, M.M.; Chen, J.L.; Fang, S.M.; Zhou, P.P. Recent advances in Cu_2O-based composites for photocatalysis: A review. *Dalton Trans.* **2021**, *50*, 4091–4111. [CrossRef] [PubMed]
25. Li, C.F.; Guo, R.T.; Zhang, Z.R.; Wu, T.; Pan, W.G. Converting CO_2 into Value-Added Products by Cu_2O-Based Catalysts: From Photocatalysis, Electrocatalysis to Photoelectrocatalysis. *Small* **2023**, *19*, 2207875. [CrossRef]

26. Yu, W.B.; Yi, M.; Fu, H.H.; Pei, M.J.; Liu, Y.; Xu, B.M.; Zhang, J. Dandelion-Like Nanostructured Cu/Cu$_2$O Heterojunctions with Fast Diffusion Channels Enabling Rapid Photocatalytic Pollutant Removal. *ACS Appl. Nano Mater.* **2023**, *6*, 2928–2941. [CrossRef]
27. Stareck, J.E. Decorating Metals. U.S. Patents 2,081,121A, 18 May 1937.
28. Zhou, Q.Q.; Chen, Y.X.; Shi, H.Y.; Chen, R.; Ji, M.H.; Li, K.X.; Wang, H.L.; Jiang, X.; Lu, C.Z. The construction of p/n-Cu$_2$O heterojunction catalysts for efficient CO$_2$ photoelectric reduction. *Catalysts* **2023**, *13*, 857. [CrossRef]
29. Sahu, K.; Bisht, A.; Khan, S.A.; Sulania, I.; Singhal, R.; Pandey, A.; Mohapatra, S. Thickness dependent optical, structural, morphological, photocatalytic and catalytic properties of radio frequency magnetron sputtered nanostructured Cu$_2$O-CuO thin films. *Ceram. Int.* **2020**, *46*, 14902–14912. [CrossRef]
30. Huang, L.; Peng, F.; Yu, H.; Wang, H. Synthesis of Cu$_2$O nanoboxes, nanocubes and nanospheres by polyol process and their adsorption characteristic. *Mater. Res. Bull.* **2008**, *43*, 3047–3053. [CrossRef]
31. Ma, L.L.; Li, J.L.; Sun, H.Z.; Qiu, M.Q.; Wang, J.B.; Chen, J.Y.; Yu, Y. Self-assembled Cu$_2$O flowerlike architecture: Polyol synthesis, photocatalytic activity and stability under simulated solar light. *Mater. Res. Bull.* **2010**, *45*, 961–968. [CrossRef]
32. Zhu, J.; Wang, Y.; Wang, X.; Yang, X.; Lu, L. A convenient method for preparing shape-controlled nanocrystalline Cu$_2$O in a polyol or water/polyol system. *Powder Technol.* **2008**, *181*, 249–254. [CrossRef]
33. Huang, X.W.; Liu, Z.J.; Zheng, Y.F. Synthesis of Cu$_2$O nanobelts via surfactant-assisted polyol method. *Chin. Chem. Lett.* **2011**, *22*, 879–882. [CrossRef]
34. Bai, Y.; Yang, T.; Gu, Q.; Cheng, G.; Zheng, R. Shape control mechanism of cuprous oxide nanoparticles in aqueous colloidal solutions. *Powder Technol.* **2012**, *227*, 35–42. [CrossRef]
35. Shin, H.S.; Song, J.Y.; Yu, J. Template-assisted electrochemical synthesis of cuprous oxide nanowires. *Mater. Lett.* **2009**, *63*, 397–399. [CrossRef]
36. Dodoo-Arhin, D.; Leoni, M.; Scardi, P.; Garnier, E.; Mittiga, A. Synthesis, characterisation and stability of Cu$_2$O nanoparticles produced via reverse micelles microemulsion. *Mater. Chem. Phys.* **2010**, *122*, 602–608. [CrossRef]
37. Wang, Y.Q.; Liang, W.S.; Satti, A.; Nikitin, K. Fabrication and microstructure of Cu$_2$O nanocubes. *J. Cryst. Growth* **2010**, *312*, 1605–1609. [CrossRef]
38. Zhang, H.; Cui, Z. Solution-phase synthesis of smaller cuprous oxide nanocubes. *Mater. Res. Bull.* **2008**, *43*, 1583–1589. [CrossRef]
39. Zhu, J.; Bi, H.; Wang, Y.; Wang, X.; Yang, X.; Lu, L. Solution-phase synthesis of Cu$_2$O cubes using CuO as a precursor. *Mater. Lett.* **2008**, *62*, 2081–2083. [CrossRef]
40. Zhang, X.; Cui, Z. One-pot growth of Cu$_2$O concave octahedron microcrystal in alkaline solution. *Mater. Sci. Eng. B Solid-State Mater. Adv. Technol.* **2009**, *162*, 82–86. [CrossRef]
41. Liang, Z.H.; Zhu, Y.J. Synthesis of uniformly sized Cu$_2$O crystals with star-like and flower-like morphologies. *Mater. Lett.* **2005**, *59*, 2423–2425. [CrossRef]
42. Yang, Y.M.; Wang, K.; Yang, Z.H.; Zhang, Y.M.; Gu, H.Y.; Zhang, W.X.; Li, E.R.; Zhou, C. An efficient route to Cu$_2$O nanorod array film for high-performance Li-ion batteries. *Thin Solid Films* **2016**, *608*, 79–87. [CrossRef]
43. Luo, Y.; Tu, Y.; Ren, Q.; Dai, X.; Xing, L.; Li, J. Surfactant-free fabrication of Cu$_2$O nanosheets from Cu colloids and their tunable optical properties. *J. Solid State Chem.* **2009**, *182*, 182–186. [CrossRef]
44. Wang, H.; He, S.; Yu, S.; Shi, T.; Jiang, S. Template-free synthesis of Cu$_2$O hollow nanospheres and their conversion into Cu hollow nanospheres. *Powder Technol.* **2009**, *193*, 182–186. [CrossRef]
45. Sui, Y.; Zhang, Y.; Fu, W.; Yang, H.; Zhao, Q.; Sun, P.; Ma, D.; Yuan, M.; Li, Y.; Zou, G. Low-temperature template-free synthesis of Cu$_2$O hollow spheres. *J. Cryst. Growth* **2009**, *311*, 2285–2290. [CrossRef]
46. Wei, M.; Huo, J. Preparation of Cu$_2$O nanorods by a simple solvothermal method. *Mater. Chem. Phys.* **2010**, *121*, 291–294. [CrossRef]
47. Wei, M.; Lun, N.; Ma, X.; Wen, S. A simple solvothermal reduction route to copper and cuprous oxide. *Mater. Lett.* **2007**, *61*, 2147–2150. [CrossRef]
48. Lim, Y.F.; Chua, C.S.; Lee, C.J.J.; Chi, D. Sol-gel deposited Cu$_2$O and CuO thin films for photocatalytic water splitting. *Phys. Chem. Chem. Phys.* **2014**, *16*, 25928–25934. [CrossRef]
49. Qu, Y.; Li, X.; Chen, G.; Zhang, H.; Chen, Y. Synthesis of Cu$_2$O nano-whiskers by a novel wet-chemical route. *Mater. Lett.* **2008**, *62*, 886–888. [CrossRef]
50. Susman, M.D.; Feldman, Y.; Vaskevich, A.; Rubinstein, I. Chemical Deposition of Cu$_2$O Nanocrystals with Precise Morphology Control. *ACS Nano* **2014**, *8*, 162–174. [CrossRef] [PubMed]
51. Balamurugan, B.; Mehta, B.R. Optical and structural properties of nanocrystalline copper oxide thin films prepared by activated reactive evaporation. *Thin Solid Films* **2001**, *396*, 90–96. [CrossRef]
52. Al-Kuhaili, M.F. Characterization of copper oxide thin films deposited by the thermal evaporation of cuprous oxide (Cu$_2$O). *Vacuum* **2008**, *82*, 623–629. [CrossRef]
53. Barreca, D.; Comini, E.; Gasparotto, A.; Maccato, C.; Sada, C.; Sberveglieri, G.; Tondello, E. Chemical vapor deposition of copper oxide films and entangled quasi-1D nanoarchitectures as innovative gas sensors. *Sens. Actuators B Chem.* **2009**, *141*, 270–275. [CrossRef]
54. Gomersall, D.E.; Flewitt, A.J. Plasma enhanced chemical vapor deposition of p-type Cu$_2$O from metal organic precursors. *J. Appl. Phys.* **2022**, *131*, 215301. [CrossRef]

55. Wang, S.; Zhang, X.; Pan, L.; Zhao, F.-M.; Zou, J.-J.; Zhang, T.; Wang, L. Controllable sonochemical synthesis of $Cu_2O/Cu_2(OH)_3NO_3$ composites toward synergy of adsorption and photocatalysis. *Appl. Catal. B Environ.* **2015**, *164*, 234–240. [CrossRef]
56. Ma, D.; Liu, H.; Yang, H.; Fu, W.; Zhang, Y.; Yuan, M.; Sun, P.; Zhou, X. High pressure hydrothermal synthesis of cuprous oxide microstructures of novel morphologies. *Mater. Chem. Phys.* **2009**, *116*, 458–463. [CrossRef]
57. Valodkar, M.; Pal, A.; Thakore, S. Synthesis and characterization of cuprous oxide dendrites: New simplified green hydrothermal route. *J. Alloys Compd.* **2011**, *509*, 523–528. [CrossRef]
58. Togashi, T.; Hitaka, H.; Ohara, S.; Naka, T.; Takami, S.; Adschiri, T. Controlled reduction of Cu^{2+} to Cu^+ with an N,O-type chelate under hydrothermal conditions to produce Cu_2O nanoparticles. *Mater. Lett.* **2010**, *64*, 1049–1051. [CrossRef]
59. Li, J.; Sun, L.; Yan, Y. One-step in-situ fabrication of silver-modified Cu_2O crystals with enhanced visible photocatalytic activity. *Micro-Nano Lett.* **2016**, *11*, 363–365. [CrossRef]
60. Wang, B.; Li, R.; Zhang, Z.; Zhang, W.; Yan, X.; Wu, X.; Cheng, G.; Zheng, R. Novel Au/CuO Multi-shelled Porous heterostructures forenhanced efficiency photoelectrochemical water splitting. *J. Mater. Chem. A* **2017**, *5*, 14415–14421. [CrossRef]
61. Jin, J.Y.; Mei, H.; Wu, H.M.; Wang, S.F.; Xia, Q.H.; Ding, Y. Selective detection of dopamine based on Cu_2O@Pt core-shell nanoparticles modified electrode in the presence of ascorbic acid and uric acid. *J. Alloy Compd.* **2016**, *689*, 174–181. [CrossRef]
62. Kuo, M.Y.; Hsiao, C.F.; Chiu, Y.H.; Lai, T.H.; Fang, M.J.; Wu, J.Y.; Chen, J.W.; Wu, C.L.; Wei, K.H.; Lin, H.C.; et al. Au@Cu_2O core@shell nanocrystals as dual-functional catalysts for sustainable environmental applications. *Appl. Catal. B* **2019**, *242*, 499–506. [CrossRef]
63. Yang, Z.; Ma, C.; Wang, W.; Zhang, M.; Hao, X.S. Chen. Long-term antibacterial stable reduced graphene oxide nanocomposites loaded with cuprous oxide nanoparticles. *J. Colloid Interface Sci.* **2019**, *557*, 156–167. [CrossRef] [PubMed]
64. Huang, Y.; Yan, C.F.; Guo, C.Q.; Lu, Z.X.; Shi, Y.; Wang, Z.D. Synthesis of GO-modified Cu_2O nanosphere and the photocatalytic mechanism of water splitting for hydrogen production. *Int. J. Hydrogen Energy* **2016**, *7*, 4007–4016. [CrossRef]
65. Yu, J.; Jin, J.; Cheng, B.; Jaroniec, M. A noble metal-free reduced graphene oxide-CdS nanorod composite for the enhanced visible-light photocatalytic reduction of CO_2 to solar fuel. *J. Mater. Chem. A* **2014**, *2*, 3407. [CrossRef]
66. Yang, L.; Luo, S.; Li, Y. High Efficient Photocatalytic Degradation of p-Nitrophenol on a Unique Cu_2O/TiO_2 p-n Heterojunction Network Catalyst. *Environ. Sci. technol.* **2010**, *44*, 7641–7646. [CrossRef]
67. Fu, J.; Cao, S.; Yu, J. Dual Z-scheme charge transfer in TiO_2-Ag-Cu_2O composite for enhanced photocatalytic hydrogen generation. *J. Mater.* **2015**, *1*, 124–133. [CrossRef]
68. Yu, L.; Ba, X.; Qiu, M.; Li, Y.; Shuai, L.; Zhang, W.; Ren, Z.; Yu, Y. Visible-light driven CO_2 reduction coupled with water oxidation on Cl-doped Cu_2O nanorods. *Nano Energy* **2019**, *60*, 576–582. [CrossRef]
69. Zhang, L.Z.; Jing, D.W.; Guo, L.J.; Yao, X.D. In situ photochemical synthesis of Zn-Doped Cu_2O hollow microcubes for high efficient photocatalytic H_2 production. *ACS Sustain. Chem. Eng.* **2014**, *2*, 1446–1452. [CrossRef]
70. Kalubowila, K.D.R.N.; Gunewardene, M.S.; Jayasingha, J.L.K.; Dissanayake, D.; Jayathilaka, C.; Jayasundara, J.M.D.; Gao, Y.; Jayanetti, J.K.D.S. Reduction-induced synthesis of reduced graphene oxide-wrapped Cu_2O/Cu nanoparticles for photodegradation of methylene blue. *ACS Appl. Nano Mater.* **2021**, *4*, 2673–2681. [CrossRef]
71. Wang, W.L.; Wang, L.; Su, W.; Xing, Y. Photocatalytic CO_2 reduction over copper-based materials: A review. *J. CO2 Util.* **2022**, *61*, 102056. [CrossRef]
72. Celaya, C.A.; Delesma, C.; Torres-Arellano, S.; Sebastian, P.J.; Muniz, J. Understanding CO_2 conversion into hydrocarbons via a photoreductive process supported on the Cu_2O(100), (110) and (111) surface facets: A first principles study. *Fuel* **2021**, *306*, 121643. [CrossRef]
73. Wu, Y.A.; McNulty, I.; Liu, C.; Lau, K.C.; Liu, Q.; Paulikas, A.P.; Sun, C.J.; Cai, Z.H.; Guest, J.R.; Ren, Y.; et al. Facet-dependent active sites of a single Cu_2O particle photocatalyst for CO_2 reduction to methanol. *Nat. Energy* **2019**, *4*, 957–968. [CrossRef]
74. Sahu, A.K.; Pokhriyal, M.; Upadhyayula, S.; Zhao, X.S. Modulating charge carrier dynamics among anisotropic crystal facets of Cu_2O for enhanced CO_2 photoreduction. *J. Phys. Chem. C* **2022**, *126*, 13094–13104. [CrossRef]
75. Zhao, Z.Y.; Yi, J.; Zhou, D.C. Electronic structures of halogen-doped Cu_2O based on DFT calculations. *Chinese. Phys. B* **2014**, *23*, 017401. [CrossRef]
76. Cheng, S.P.; Wei, L.W.; Wang, H.P. Photocatalytic reduction of CO_2 to methanol by Cu_2O/TiO_2 heterojunctions. *Sustainability* **2022**, *14*, 374. [CrossRef]
77. Zhang, F.; Li, Y.H.; Qi, M.Y.; Tang, Z.R.; Xu, Y.J. Boosting the activity and stability of Ag-Cu_2O/ZnO nanorods for photocatalytic CO_2 reduction. *Appl. Catal. B-Environ.* **2020**, *268*, 118380. [CrossRef]
78. Song, Y.Y.; Zhao, X.J.; Feng, X.Y.; Chen, L.M.; Yuan, T.C.; Zhang, F.Q. Z-scheme $Cu_2O/Cu/Cu_3V_2O_7(OH)_2 \cdot 2H_2O$ heterostructures for efficient visible-light photocatalytic CO_2 reduction. *ACS Appl. Energy Mater.* **2022**, *5*, 10542–10552. [CrossRef]
79. Cui, L.K.; Hu, L.Q.; Shen, Q.Q.; Liu, X.G.; Jia, H.S.; Xue, J.B. Three-dimensional porous Cu_2O with dendrite for efficient photocatalytic reduction of CO_2 under visible light. *Appl. Catal. B-Environ.* **2022**, *581*, 152343. [CrossRef]
80. Zheng, Y.K.; Duan, Z.T.; Liang, R.X.; Lv, R.Q.; Wang, C.; Zhang, Z.X.; Wan, S.L.; Wang, S.; Xiong, H.F.; Ngaw, C.K. Shape-dependent performance of Cu/Cu_2O for photocatalytic reduction of CO_2. *ChemSusChem* **2022**, *15*, e202200216. [CrossRef]
81. Zhang, X.J.; Zhao, X.; Chen, K.; Fan, Y.Y.; Wei, S.L.; Zhang, W.S.; Han, D.X.; Niu, L. Palladium-modified cuprous(i) oxide with {100} facets for photocatalytic CO_2 reduction. *Nanoscale* **2021**, *13*, 2883–2890. [CrossRef]

82. Zhao, X.X.; Sun, L.L.; Jin, X.; Xu, M.Y.; Yin, S.K.; Li, J.Z.; Li, X.; Shen, D.; Yan, Y.; Huo, P.W. Cu media constructed Z-scheme heterojunction of UiO-66-NH$_2$/Cu$_2$O/Cu for enhanced photocatalytic induction of CO$_2$. *Appl. Surf. Sci.* **2021**, *545*, 148967. [CrossRef]
83. Deng, Y.; Wan, C.A.; Li, C.; Wang, Y.Y.; Mu, X.Y.; Liu, W.; Huang, Y.P.; Wong, P.K.; Ye, L.Q. Synergy effect between facet and zero-valent copper for selectivity photocatalytic methane formation from CO$_2$. *ACS Catal.* **2022**, *12*, 4526–4533. [CrossRef]
84. Tang, Z.L.; He, W.J.; Wang, Y.L.; Wei, Y.C.; Yu, X.L.; Xiong, J.; Wang, X.; Zhang, X.; Zhao, Z.; Liu, J. Ternary heterojunction in rGO-coated Ag/Cu$_2$O catalysts for boosting selective photocatalytic CO$_2$ reduction into CH$_4$. *Appl. Catal. B-Environ.* **2022**, *311*, 121371. [CrossRef]
85. Zhou, J.Q.; Li, Y.F.; Yu, L.; Li, Z.P.; Xie, D.F.; Zhao, Y.Y.; Yu, Y. Facile in situ fabrication of Cu$_2$O@Cu metal-semiconductor heterostructured nanorods for efficient visible-light driven CO$_2$ reduction. *Chem. Eng. J.* **2020**, *385*, 123940. [CrossRef]
86. Ali, S.; Lee, J.; Kim, H.; Hwang, Y.; Razzaq, A.; Jung, J.W.; Cho, C.H.; In, S.I. Sustained, photocatalytic CO$_2$ reduction to CH$_4$ in a continuous flow reactor by earth-abundant materials: Reduced titania-Cu$_2$O Z-scheme heterostructures. *Appl. Catal. B-Environ.* **2020**, *279*, 119344. [CrossRef]
87. Yao, S.; Sun, B.Q.; Zhang, P.; Tian, Z.Y.; Yin, H.Q.; Zhang, Z.M. Anchoring ultrafine Cu$_2$O nanocluster on PCN for CO$_2$ photoreduction in water vapor with much improved stability. *Appl. Catal. B-Environ.* **2022**, *317*, 121702. [CrossRef]
88. Movahed, S.K.; Najinasab, A.; Nikbakht, R.; Dabiri, M. Visible light assisted photocatalytic reduction of CO$_2$ to methanol using Fe$_3$O$_4$@N-C/Cu$_2$O nanostructure photocatalyst. *J. Photochem. Photobiol. A* **2020**, *401*, 112763. [CrossRef]
89. Liu, S.H.; Lu, J.S.; Pu, Y.C.; Fan, H.C. Enhanced photoreduction of CO$_2$ into methanol by facet-dependent Cu$_2$O/reduce graphene oxide. *J. CO2 Util.* **2019**, *33*, 171–178. [CrossRef]
90. Li, H.T.; Deng, Y.D.; Liu, Y.D.; Zeng, X.; Wiley, D.; Huang, J. Carbon quantum dots and carbon layer double protected cuprous oxide for efficient visible light CO$_2$ reduction. *Chem. Commun.* **2019**, *55*, 4419–4422. [CrossRef]
91. Zeng, Z.P.; Yan, Y.B.; Chen, J.; Zan, P.; Tian, Q.H.; Chen, P. Boosting the photocatalytic ability of Cu$_2$O nanowires for CO$_2$ conversion by mxene quantum dots. *Adv. Funct. Mater.* **2018**, *29*, 1806500. [CrossRef]
92. Wang, H.; Cheng, S.J.; Cai, X.; Cheng, L.H.; Zhou, R.J.; Hou, T.T.; Li, Y.W. Photocatalytic CO$_2$ reduction to HCOOH over core-shell Cu@Cu$_2$O catalysts. *Catal. Commun.* **2022**, *162*, 106372. [CrossRef]
93. Zhu, Q.; Cao, Y.N.; Tao, Y.; Li, T.; Zhang, Y.; Shang, H.; Song, J.X.; Li, G.S. CO$_2$ reduction to formic acid via NH$_2$-C@Cu$_2$O photocatalyst in situ derived from amino modified Cu-MOF. *J. CO2 Util.* **2021**, *54*, 101781. [CrossRef]
94. Dominguez-Arvizu, J.L.; Jimenez-Miramontes, J.A.; Hernandez-Majalca, B.C.; Valenzuela-Castro, G.E.; Gaxiola-Cebreros, F.A.; Salinas-Gutierrez, J.M.; Collins-Martinez, V.; Lopez-Ortiz, A. Study of NiFe$_2$O$_4$/Cu$_2$O p-n heterojunctions for hydrogen production by photocatalytic water splitting with visible light. *J. Mater. Res. Technol.* **2023**, *21*, 4184–4199. [CrossRef]
95. Yang, J.F.; Yang, H.Y.; Dong, Y.H.; Cui, H.; Sun, H.; Yin, S.Y. Fabrication of Cu$_2$O/MTiO$_3$ (M = Ca, Sr and Ba) p-n heterojunction for highly enhanced photocatalytic hydrogen generation. *J. Alloy Compd.* **2023**, *930*, 167333. [CrossRef]
96. Zhang, Y.H.; Liu, M.M.; Chen, J.L.; Xie, K.F.; Fang, S.M. Dendritic branching Z-scheme Cu$_2$O/TiO$_2$ heterostructure photocatalysts for boosting H$_2$ production. *J. Phys. Chem. Solids* **2021**, *152*, 109948. [CrossRef]
97. Dai, B.L.; Li, Y.Y.; Xu, J.M.; Sun, C.; Li, S.J.; Zhao, W. Photocatalytic oxidation of tetracycline, reduction of hexavalent chromium and hydrogen evolution by Cu$_2$O/g-C$_3$N$_4$ S-scheme photocatalyst: Performance and mechanism insight. *Appl. Surf. Sci.* **2022**, *592*, 153309. [CrossRef]
98. Liu, Y.X.; Tan, H.; Wei, Y.A.; Liu, M.H.; Hong, J.X.; Gao, W.Q.; Zhao, S.Q.; Zhang, S.P.; Guo, S.J. Cu$_2$O/2D COFs core/shell nanocubes with antiphotocorrosion ability for efficient evolution. *ACS Nano* **2023**, *17*, 5994–6001. [CrossRef]
99. Muscetta, M.; Al Jitan, S.; Palmisano, G.; Andreozzi, R.; Marotta, R.; Cimino, S.; Di Somma, I. Visible light-driven photocatalytic hydrogen production using Cu$_2$O/TiO$_2$ composites prepared by facile mechanochemical synthesis. *J. Environ. Chem. Eng.* **2022**, *10*, 107735. [CrossRef]
100. Chang, Y.C.; Syu, S.Y.; Lu, M.Y. Fabrication of In(OH)$_3$-In$_2$S$_3$-Cu$_2$O nanofiber for highly efficient photocatalytic hydrogen evolution under blue light LED excitation. *Int. J. Hydrogen Energy* **2023**, *48*, 9318–9332. [CrossRef]
101. Lai, T.H.; Tsao, C.W.; Fang, M.J.; Wu, J.Y.; Chang, Y.P.; Chiu, Y.H.; Hsieh, P.Y.; Kuo, M.Y.; Chang, K.D.; Hsu, Y.J. Au@Cu$_2$O core-shell and Au@Cu$_2$Se yolk-shell nanocrystals as promising photocatalysts in photoelectrochemical water splitting and photocatalytic hydrogen production. *ACS Appl. Mater. Interfaces* **2022**, *14*, 40771–40783. [CrossRef]
102. Ibarra-Rodriguez, L.I.; Huerta-Flores, A.M.; Torres-Martinez, L.M. Development of Na$_2$Ti$_6$O$_{13}$/CuO/Cu$_2$O heterostructures for solar photocatalytic production of low-carbon fuels. *Mater. Res. Bull.* **2020**, *122*, 110679. [CrossRef]
103. Dubale, A.A.; Ahmed, I.N.; Zhang, Y.J.; Yang, X.L.; Xie, M.H. A facile strategy for fabricating C@Cu$_2$O/CuO composite for efficient photochemical hydrogen production with high external quantum efficiency. *Appl. Surf. Sci.* **2020**, *534*, 147582. [CrossRef]
104. Fan, Z.B.; Zhang, X.J.; Li, Y.J.; Guo, X.; Jin, Z.L. Construct 3D NiCo-LDH/Cu$_2$O p-n heterojunction via electrostatic self-assembly for enhanced photocatalytic hydrogen evolution. *J. Ind. Eng. Chem.* **2022**, *110*, 491–502. [CrossRef]
105. Mohite, S.V.; Kim, S.; Lee, C.S.; Bae, J.; Kim, Y. Z-scheme heterojunction photocatalyst: Deep eutectic solvents-assisted synthesis of Cu$_2$O nanocluster improved hydrogen production of TiO$_2$. *J. Alloy Compd.* **2022**, *928*, 167168. [CrossRef]
106. Qiu, P.; Xiong, J.Y.; Lu, M.J.; Liu, L.J.; Li, W.; Wen, Z.P.; Li, W.J.; Chen, R.; Cheng, G. Integrated p-n/Schottky junctions for efficient photocatalytic hydrogen evolution upon Cu@TiO$_2$-Cu$_2$O ternary hybrids with steering charge transfer. *J. Colloid Interface Sci.* **2022**, *622*, 924–937. [CrossRef] [PubMed]

107. Becerra-Paniagua, D.K.; Torres-Arellano, S.; Martinez-Alonso, C.; Luevano-Hipolito, E.; Sebastian, P.J. Facile and green synthesis of Cu/Cu$_2$O composite for photocatalytic H$_2$ generation. *Mater. Sci. Semicond. Process.* **2023**, *162*, 107485. [CrossRef]
108. Zhu, H.; Xi, M.Y.; Huang, G.P.; Qin, L.X.; Zhang, T.Y.; Kang, S.Z.; Li, X.Q. Cuprous oxide core-shell heterostructure facilely encapsulated by cadmium metal organic frameworks for enhanced photocatalytic hydrogen generation. *J. Phys. Chem. Solids* **2023**, *181*, 111476. [CrossRef]
109. Zhang, Y.; Li, X.J.; Chen, J.F.; Wang, Y.A.; Cheng, Z.Y.; Chen, X.Q.; Gao, X.; Guo, M.H. Porous spherical Cu$_2$O supported by wood-based biochar skeleton for the adsorption-photocatalytic degradation of methyl orange. *Appl. Surf. Sci.* **2023**, *611*, 155744. [CrossRef]
110. Sehrawat, P.; Rana, S.; Mehta, S.K.; Kansal, S.K. Optimal synthesis of MoS$_2$/Cu$_2$O nanocomposite to enhance photocatalytic performance towards indigo carmine dye degradation. *Appl. Surf. Sci.* **2022**, *604*, 154482. [CrossRef]
111. Li, J.J.; Guo, C.P.; Li, L.H.; Gu, Y.J.; BoK-Hee, K.; Huang, J.L. Construction of Z-scheme WO$_3$-Cu$_2$O nanorods array heterojunction for efficient photocatalytic degradation of methylene blue. *Inorg. Chem. Commun.* **2022**, *138*, 109248. [CrossRef]
112. Nie, J.K.; Yu, X.J.; Liu, Z.B.; Zhang, J.; Ma, Y.; Chen, Y.Y.; Ji, Q.G.; Zhao, N.N.; Chang, Z. Energy band reconstruction mechanism of Cl-doped Cu$_2$O and photocatalytic degradation pathway for levofloxacin. *J. Clean. Prod.* **2022**, *363*, 132593. [CrossRef]
113. Wu, Y.; Li, Y.Q.; Li, H.; Guo, H.; Yang, Q.; Li, X.M. Tunning heterostructures interface of Cu$_2$O@HKUST-1 for enhanced photocatalytic degradation of tetracycline hydrochloride. *Sep. Purif. Technil.* **2022**, *303*, 122106. [CrossRef]
114. Xiong, J.; Zeng, H.Y.; Peng, J.F.; Peng, D.Y.; Liu, F.Y.; Xu, S.; Yang, Z.L. abrication of Cu$_2$O/ZnTi-LDH p-n heterostructure by grafting Cu$_2$O NPs onto the LDH host layers from Cu-doped ZnTi-LDH and insight into the photocatalytic mechanism. *Compos. Part. B-Eng.* **2023**, *250*, 110447. [CrossRef]
115. Zhu, P.; Li, Y.; Ma, Y.; Ruan, X.X.; Zhang, Q.Z. Preparation of waste-based Cu-Cu$_2$O/SiO$_2$ photocatalyst from serpentine tailings and waste printed circuit boards and photoreduction of Cr(VI). *Ceram. Int.* **2023**, *49*, 12518–12528. [CrossRef]
116. Alp, E. The facile synthesis of Cu$_2$O-Cu hybrid cubes as efficient visible-light-driven photocatalysts for water remediation processes. *Powder Technol.* **2021**, *394*, 1111–1120. [CrossRef]
117. Mkhalid, I.A.; Shawky, A. Cu-supported Cu$_2$O nanoparticles: Optimized photodeposition enhances the visible light photodestruction of atrazine. *J. Alloy Compd.* **2021**, *853*, 157040. [CrossRef]
118. Dubale, A.A.; Ahmed, I.N.; Chen, X.H.; Ding, C.; Hou, G.H.; Guan, R.F.; Meng, X.M.; Yang, X.L.; Xie, M.H. A highly stable metal-organic framework derived phosphorus doped carbon/Cu$_2$O structure for efficient photocatalytic phenol degradation and hydrogen production. *J. Mater. Chem. A* **2019**, *7*, 6062–6079. [CrossRef]
119. Yu, X.J.; Zhang, J.; Chen, Y.Y.; Ji, Q.G.; Wei, Y.C.; Niu, J.N.; Yu, Z.; Yao, B.H. Ag-Cu$_2$O composite films with enhanced photocatalytic activities for methylene blue degradation: Analysis of the mechanism and the degradation pathways. *J. Environ. Chem. Eng.* **2021**, *9*, 106161. [CrossRef]
120. Nie, J.K.; Yu, X.J.; Liu, Z.B.; Wei, Y.C.; Zhang, J.; Zhao, N.N.; Yu, Z.; Yao, B.H. Boosting principles for the photocatalytic performance of Cr-doped Cu$_2$O crystallites and mechanisms of photocatalytic oxidation for levofloxacin. *Appl. Surf. Sci.* **2021**, *576*, 151842. [CrossRef]
121. Liu, Z.B.; Yu, X.J.; Gao, P.H.; Nie, J.K.; Yang, F.; Guo, B.Q.; Zhang, J. Preparation of BiOCl/Cu$_2$O composite particles and its photocatalytic degradation of moxifloxacin. *Opt. Mater.* **2022**, *128*, 112432. [CrossRef]
122. Zhang, Y.Y.; Li, Y.; Ruan, Z.H.; Yuan, Y.; Lin, K.F. Extensive solar light utilizing by ternary C-dots/Cu$_2$O/SrTiO$_3$: Highly enhanced photocatalytic degradation of antibiotics and inactivation of E. coli. *Chemosphere* **2022**, *290*, 133340. [CrossRef]
123. Bayat, F.; Sheibani, S. Enhancement of photocatalytic activity of CuO-Cu$_2$O heterostructures through the controlled content of Cu$_2$O. *Mater. Res. Bull.* **2021**, *145*, 111561. [CrossRef]
124. Su, X.P.; Chen, W.; Han, Y.N.; Wang, D.C.; Yao, J.M. In-situ synthesis of Cu$_2$O on cotton fibers with antibacterial properties and reusable photocatalytic degradation of dyes. *Appl. Surf. Sci.* **2021**, *536*, 147945. [CrossRef]
125. Wu, Y.; Li, X.M.; Zhao, H.; Yao, F.B.; Cao, J.; Chen, Z.; Wang, D.B.; Yang, Q. Core-shell structured Cu$_2$O@HKUST-1 heterojunction photocatalyst with robust stability for highly efficient tetracycline hydrochloride degradation under visible light. *Chem. Eng. J.* **2021**, *426*, 131255. [CrossRef]
126. Huang, J.; Zhou, T.X.; Zhao, W.S.; Cui, S.C.; Guo, R.; Li, D.; Kadasala, N.R.; Han, D.L.; Jiang, Y.H.; Liu, Y. Multifunctional magnetic Fe$_3$O$_4$/Cu$_2$O-Ag nanocomposites with high sensitivity for SERS detection and efficient visible light-driven photocatalytic degradation of polycyclic aromatic hydrocarbons (PAHs). *J. Colloid Interf. Sci.* **2022**, *628*, 315–326. [CrossRef]
127. Altynbaeva, L.S.; Barsbay, M.; Aimanova, N.A.; Jakupova, Z.Y.; Nurpeisova, D.T.; Zdorovets, M.V.; Mashentseva, A.A. A Novel Cu$_2$O/ZnO@PET Composite Membrane for the Photocatalytic Degradation of Carbendazim. *Nanomaterials* **2022**, *12*, 1724. [CrossRef]
128. Yanagida, S.; Yajima, T.; Takei, T.; Kumada, N. Removal of hexavalent chromium from water by Z-scheme photocatalysis using TiO$_2$ (rutile) nanorods loaded with Au core-Cu$_2$O shell particles. *J. Environ. Sci.* **2022**, *115*, 173–189. [CrossRef] [PubMed]
129. Qiang, T.T.; Wang, S.T.; Ren, L.F.; Gao, X.D. Novel 3D Cu$_2$O/N-CQD/ZIF-8 composite photocatalyst with Z-scheme heterojunction for the efficient photocatalytic reduction of Cr(VI). *J. Environ. Chem. Eng.* **2022**, *10*, 108784. [CrossRef]
130. Ma, J.; Liang, C.J.; Yu, C.J.; Li, H.M.; Xu, H.; Hua, Y.J.; Wang, C.T. BiOBr microspheres anchored with Cu$_2$O nanoparticles and rGO: A Z-scheme heterojunction photocatalyst for efficient reduction of Cr(VI) under visible light irradiation. *Appl. Surf. Sci.* **2023**, *609*, 155247. [CrossRef]

131. An, X.Q.; Liu, H.J.; Qu, J.H.; Moniz, S.J.A.; Tang, J.W. Photocatalytic mineralisation of herbicide 2,4,5-trichlorophenoxyacetic acid: Enhanced performance by triple junction Cu-TiO$_2$-Cu$_2$O and the underlying reaction mechanism. *New J. Chem.* **2015**, *39*, 314–320. [CrossRef]
132. Sharma, K.; Maiti, K.; Kim, N.H.; Hui, D.; Lee, J.H. Green synthesis of glucose-reduced graphene oxide supported Ag-Cu$_2$O nanocomposites for the enhanced visible-light photocatalytic activity. *Compos. Part. B-Eng.* **2018**, *138*, 35–44. [CrossRef]
133. Lv, Q.Y.; Li, L.L.; Li, Y.F.; Mao, J.H.; Chen, T.; Shao, D.W.; Li, M.M.; Tan, R.S.; Zhao, J.Q.; Shi, S.H. A DFT Study of Electronic Structures and Photocatalytic Properties of Mn-Cu$_2$O. *Russ. J. Phys. Chem. A* **2020**, *94*, 641–646.

Disclaimer/Publisher's Note: The statements, opinions and data contained in all publications are solely those of the individual author(s) and contributor(s) and not of MDPI and/or the editor(s). MDPI and/or the editor(s) disclaim responsibility for any injury to people or property resulting from any ideas, methods, instructions or products referred to in the content.

Communication

Synthesis, Structural Characterization, Hirschfeld Surface Analysis and Photocatalytic CO_2 Reduction Activity of a New Dinuclear Gd(III) Complex with 6-Phenylpyridine-2-Carboxylic Acid and 1,10-Phenanthroline Ligands

Li-Hua Wang [1] and Xi-Shi Tai [2,*]

[1] College of Biology and Oceanography, Weifang University, Weifang 261061, China
[2] College of Chemistry and Chemical Engineering, Weifang University, Weifang 261061, China
* Correspondence: taixs@wfu.edu.cn; Tel.: +86-536-8785282; Fax: +86-536-8785286

Abstract: A new dinuclear Gd(III) complex was synthesized and named $[Gd_2(L)_4(Phen)_2(H_2O)_2(DMF)_2]$ ·$2H_2O$·$2Cl$ (**1**). Here, L is the 6-phenylpyridine-2-carboxylate anion, Phen represents 1,10-phenanthroline, DMF is called *N*,*N*-dimethylformamide, and Cl^- is the chloride anion, which is characterized by IR and single crystal X-ray diffraction analysis. The structural analysis reveals that complex (**1**) is a cation–anion complex, and each Gd(III) ion is eight-coordinated with four O atoms (O1, O5, O2a, O4a, or O1a, O2, O4, O5a) of four different bidentate L ligands, two O atoms (O6, or O6a) of DMF molecules, two N atoms (N1, N2, or N1a, N2a) of Phen ligands, and two O atoms (O3 or O3a) of coordinated water molecules. Complex (**1**) forms the three-dimensional π–π stacking network structure with cavities occupied by chloride anions and uncoordinated water molecules. The Hirschfeld surface of the complex (**1**) shows that the H···H contacts represented the largest contribution (48.5%) to the Hirschfeld surface, followed by C···H/H···C and O···H/H···O contacts with contributions of 27.2% and 6.0%, respectively. To understand the electronic structure of the complex (**1**), the DFT calculations have been performed. The photocatalytic CO_2 reduction activity shows complex (**1**) has excellent catalytic activity with yields of 22.1 μmol/g (CO) and 6.0 μmol/g (CH_4) after three hours. And the selectivity of CO can achieve 78.5%.

Keywords: 6-phenylpyridine-2-carboxylic acid; 1,10-phenanthroline; dinuclear Gd(III) complex; synthesis; crystal structure; hirschfeld surface analysis; photocatalytic CO_2 reduction

Citation: Wang, L.-H.; Tai, X.-S. Synthesis, Structural Characterization, Hirschfeld Surface Analysis and Photocatalytic CO_2 Reduction Activity of a New Dinuclear Gd(III) Complex with 6-Phenylpyridine-2-Carboxylic Acid and 1,10-Phenanthroline Ligands. *Molecules* **2023**, *28*, 7595. https://doi.org/10.3390/molecules28227595

Academic Editor: Lin Ju

Received: 22 October 2023
Revised: 10 November 2023
Accepted: 10 November 2023
Published: 14 November 2023

Copyright: © 2023 by the authors. Licensee MDPI, Basel, Switzerland. This article is an open access article distributed under the terms and conditions of the Creative Commons Attribution (CC BY) license (https://creativecommons.org/licenses/by/4.0/).

1. Introduction

The increased human social activities not only accelerate the consumption of fossil energy but also result in a surge in CO_2 concentration in the atmosphere. Therefore, the catalytic conversion and utilization of CO_2 have become a research hotspot. At present, studies on photocatalytic CO_2 reduction have received extensive attention [1]. So far, many studies have reported that precious metal catalysts show high catalytic activity and selectivity in the photocatalytic CO_2 reduction reaction [2–7]. However, the disadvantages of expensive costs and a lack of storage limit the further applications of precious metal catalysts. It is a trend to explore new-type photocatalysts. Recently, some metal complex photocatalysts have become one of the research hotspots due to their excellent properties in photocatalytic CO_2 reduction [8–14]. However, their activities are still low and cannot meet the needs of industrial applications. Rare earth elements often exhibit many special activities due to their special electronic structures. Therefore, earth-based rare complex photocatalysts are likely to exhibit good activity in photocatalytic CO_2 reduction. And there are few reports on THE photocatalytic CO_2 reduction in rare earth metal complexes [15,16].

Herein, we have chosen the Gd elements as the research object and prepared a new dinuclear Gd(III) complex using $GdCl_3 \cdot 6H_2O$, 6-phenylpyridine-2-carboxylic acid, 1,10-phenanthroline, and NaOH as reactants. The complex (**1**) was characterized by IR and single crystal X-ray diffraction analysis, thereby confirming it is a cation–anion complex, and each Gd(III) ion is eight-coordinated with four O atoms. Subsequently, the photocatalytic CO_2 reduction activity of complex (**1**) has been explored and found to have excellent catalytic activity, with yields of 22.1 μmol/g (CO) and 6.0 μmol/g (CH_4) after three hours. And the selectivity of CO can achieve 78.5%. The synthetic route for complex (**1**) is shown in Scheme S1 (Supplementary Materials).

2. Results and Discussion

2.1. Infrared Spectra

The infrared spectrum of complex (**1**) is given in Figure 1. The 6-phenylpyridine-2-carboxylic acid ligand exhibited characteristic bands at ca. 1646 ($\nu_{as}COO^-$) and 1575 ($\nu_s COO^-$) cm^{-1} [17], and in complex (**1**), they appeared at ca. 1619 ($\nu_{as}COO^-$), and 1425 ($\nu_s COO^-$) cm^{-1}, respectively. Phen ligand showed characteristic bands at 1597 (C=N) cm^{-1} [18], and in complex (**1**), it appeared at 1577 (C=N) cm^{-1}. These results indicate that the 6-phenylpyridine-2-carboxylic acid ligand and Phen ligand are coordinated with the Gd(III) ion. The IR results are consistent with single-crystal X-ray diffraction measurements of complex (**1**).

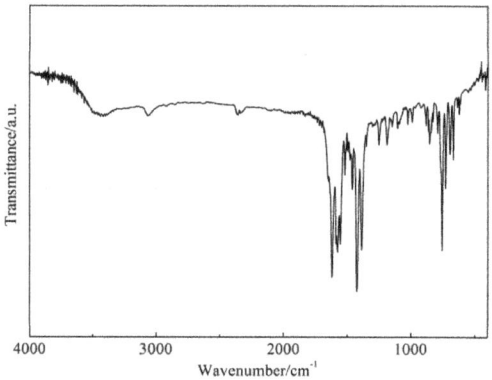

Figure 1. The infrared spectrum of the complex (**1**).

2.2. Structural Description of Complex (**1**)

The coordination environment of Gd(III) in complex (**1**) is given in Figure 2. Selected bond lengths (Å) and angles (°) for complex (**1**) are given in Table 1. Figure 3 shows the 3D network structure of complex (**1**). The asymmetric unit of complex (**1**) contains one Gd(III) ion, two 6-phenylpyridine-2-carboxylate ligands, one 1,10-phenanthroline ligand, one coordinated N,N-dimethylformamide molecule, one coordinated water molecule, one uncoordinated chloride anion, and one uncoordinated water molecule (Figure 2). The structural analysis reveals that complex (**1**) is a cation–anion complex and two Gd(III) ions are eight-coordinated with four O atoms (O1, O5, O2a, O4a, or O1a, O2, O4, O5a) of four different bidentate L ligands, one O atom (O6, or O6a) of DMF molecules, two N atoms (N1, N2, or N1a, N2a) of Phen ligands, and one O atom (O3 or O3a) of coordinated water molecules (symmetry code: 0.5-x, 1.5-y, 1.5-z). Complex (**1**) forms a dinuclear structure by bidentate chelate coordination mode of 6-phenylpyridine-2-carboxylate ligands, and the distance of two adjacent Gd(III) ions is 4.388 Å. Eight oxygen atoms are coordinated with two adjacent Gd(III) ions to form two stable eight-membered rings: ring 1 (O1-Gd1-O2a-C36a-O1a-Gd1a-O2-C36-O1) and ring 2 (O4a-Gd1-O5-C24-O4-Gd1a-O5a-C24a-O4a). The dihedral angle of ring one and ring two is 87.32°, indicating that the two eight-membered rings are nearly vertical. The bond distances of Gd–O and Gd–N are 2.313(3) Å (Gd1-O1), 2.379(3) Å (Gd1-O2a), 2.436(3) Å (Gd1-O3), 2.372(3) Å (Gd1-O4a), 2.331(3) Å (Gd1-O5), 2.406(3) Å (Gd1-O6), 2.560(3) Å (Gd1-N1), and 2.585(3) Å (Gd1-N2), respectively, which are consistent with those reported in the literature [14,19,20]. The uncoordinated chloride anion and the uncoordinated water molecule are embedded in the molecule by intramolecular

O-H···Cl hydrogen bonds. And complex (**1**) molecules form a three-dimensional network structure (Figure 3) by the π–π interaction of aromatic rings.

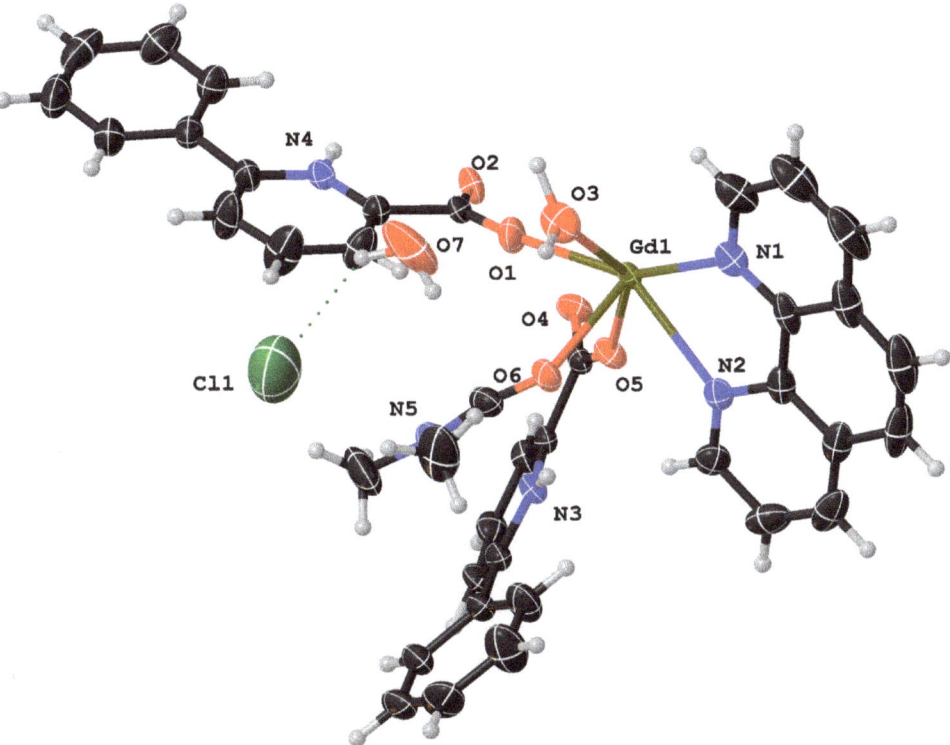

Figure 2. The coordination environment of Gd(III) in complex (**1**).

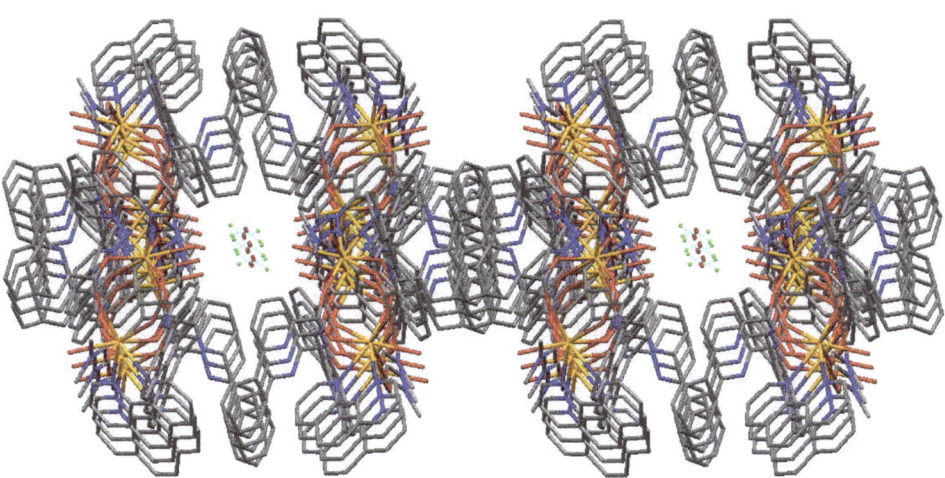

Figure 3. Three-dimensional network structure of the complex (**1**).

Table 1. Selected bond lengths (Å) and bond angles (°) for complex (**1**).

Bond	d	Angle	(°)
Gd1-N1	2.560(3)	N1-Gd1-N2	63.79(11)
Gd1-N2	2.585(3)	N1-Gd1-O1	147.23(12)
Gd1-O1	2.313(3)	O1-Gd1-N2	145.07(11)
Gd1-O2a	2.379(3)	O1-Gd1-O2a	123.07(11)
Gd1-O3	2.436(3)	O1-Gd1-O3	77.17(12)
Gd1-O4a	2.372(3)	O1-Gd1-O4a	79.20(10)
Gd1-O5	2.331(3)	O1-Gd1-O5	74.45(11)
Gd1-O6	2.406(3)	O1-Gd1-O6	81.61(11)
		N1-Gd1-O2a	74.76(11)
		O2a-Gd1-N2	69.35(10)
		O2a-Gd1-O3	139.53(10)
		O2a-Gd1-O6	139.90(10)
		N1-Gd1-O3	73.26(12)
		O3-Gd1-N2	115.87(12)
		N1-Gd1-O4a	79.48(11)
		O4a-Gd1-N2	134.52(10)
		O4a-Gd1-O2a	76.06(10)
		O4a-Gd1-O3	74.29(11)
		O4a-Gd1-O6	143.58(10)
		N1-Gd1-O5	138.32(11)
		N2-Gd1-O5	77.15(11)
		O5-Gd1-O2a	78.50(10)
		O5-Gd1-O3	141.59(11)
		O4a-Gd1-O5	123.93(10)
		O5-Gd1-O6	79.25(10)
		O6-Gd1-N1	101.73(11)
		N2-Gd1-O6	73.42(10)
		O6-Gd1-O3	71.41(10)

Symmetry transformations: a: $1/2 - x, -1 + y, 1 - z$.

Hydrogen bonds and π–π interaction play important roles in forming the 3D supermolecule of complex (**1**), and the detailed parameters are listed in the following Tables 2 and 3.

Table 2. Detailed parameters of hydrogen bonds in complex (**1**).

Donor-H	Acceptor	D-H (Å)	H⋯A (Å)	D⋯A (Å)	D-H⋯A (°)
O3-H3A	O7	1.07	2.46	3.125(6)	120
O3-H3B	O7 [#1]	1.07	2.59	3.294(6)	123
O7-H7B	Cl1	0.85	2.50	3.143(5)	133

Symmetric operation code: [#1]: $1/2 - x, y, 1 - z$.

Table 3. Detailed parameters of π–π stacking interactions in complex (**1**).

Ring1	Ring2	Symmetry	Distance between Ring Centroids	Slippage
Cg2	Cg8		3.713(2)	1.412
Cg5	Cg5		3.465(3)	0.620
Cg5	Cg8		3.746(3)	1.563
Cg5	Cg9		3.432(3)	0.583
Cg5	Cg10		3.464(2)	0.731
Cg8	Cg2	$1/2 - x, 1/2 - y, 3/2 - z$	3.713(2)	1.577
Cg8	Cg5		3.746(3)	1.554
Cg8	Cg9		3.526(2)	1.002
Cg9	Cg5		3.432(3)	0.527
Cg9	Cg8		3.526(2)	0.963
Cg9	Cg9		3.800(2)	1.722

Table 3. *Cont.*

Ring1	Ring2	Symmetry	Distance between Ring Centroids	Slippage
Cg9	Cg10		3.586(2)	1.172
Cg10	Cg5	$1/2 - x, 1/2 - y, 3/2 - z$	3.465(2)	0.711
Cg10	Cg9		3.586(2)	1.186
Cg10	Cg10		3.6798(19)	1.445

Ring number: Cg2: N2-C6-C7-C10-C11-C12; Cg5: C4-C5-C6-C7-C8-C9; Cg8: N1-C1-C2-C3-C4-C9-C8-C7-C6-C5; Cg9: N2-C6-C5-C4-C9-C8-C7-C10-C11-C12; Cg10: N1-C1-C2-C3-C4-C9-C8-C7-C10-C11-C12-N2-C6-C5.

2.3. DFT Computation

In order to understand the electronic structure of the complex, DFT calculations were performed. Figure 4 shows the electron density distributions and energy levels (eV) of HOMO-1, HOMO, LUMO, and LUMO + 1 for the ligands L and phen. These two ligands, neutral 6-phenylpyridine-2-carboxylic acid (L) and phen, were optimized at the theoretical level of B3LYP/6–31G* with the Gaussian 16 package [21–23]. In contrast to the planar phen, there is a dihedral angle of 15.5^0 between the phenyl group and pyridine subunit in the ligand L, which is different from those of the ligands L in the complex (**1**) in the crystal (27.4^0 and 8.9^0). It indicates that the coordinates and steric hindrance in complex (**1**) change the planarity of the ligand L. Moreover, the electron density distributions and energy levels of the frontier molecular orbitals are shown in Figure 4, which were realized by the VMD package and the Multiwfn program [24].

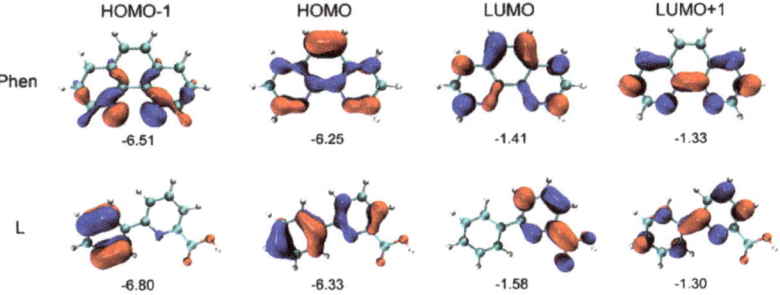

Figure 4. Electron density distributions and energy levels (eV) of HOMO-1, HOMO, LUMO, and LUMO+1 for the ligands L and phen (isovalue = 0.05 e•bohr-3). All energies are in eV.

2.4. Hirschfeld Surface Analysis of Complex (1)

The Hirschfeld surface of the complex (**1**) was analyzed by the CrystalExplorer software. As shown in Figure 5, the original crystal structure unit, the Hirschfeld surfaces, are mapped over the dnorm, di, and de of the crystal (Figure 5a–d). The two-dimensional (2D) fingerprint plots represented overall, and the top three interactions (H···H, C···H/H···C, and O···H/H···O) were shown in (Figure 5e–h). Based on the calculations, it can be concluded that the H···H contacts represented the largest contribution (48.5%) to the Hirschfeld surface, followed by C···H/H···C and O···H/H···O contacts with contributions of 27.2% and 6.0%, respectively. It is worth noting that the π–π stacking interactions play a subordinate role in forming the crystal for the C···C contacts with a Hirschfeld surface contribution percentage of 5.1%.

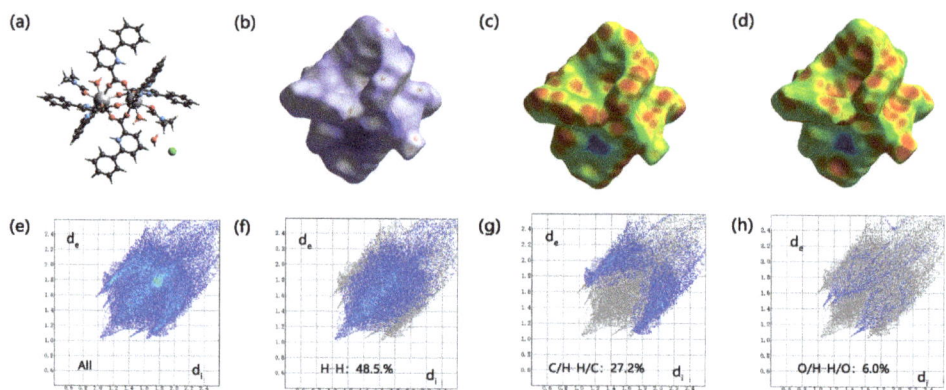

Figure 5. The Hirschfeld surface of the complex (**1**).

2.5. Photocatalytic CO_2 Reduction Activity Assessment of the Complex (**1**)

As a newly discovered Gd(III) complex material, we want to try to explore its application field. Therefore, we tested its photocatalytic CO_2 reduction activity, and the results are shown in Figure 6. Figure 6(a) shows that photocatalytic CO_2 reduction activity using complex (**1**). Figure 6(b) describes that product selectivity diagram in the photocatalytic CO_2 reduction reaction. It can be observed that the Gd(III) complex exhibits obvious photocatalytic CO_2 activity. The main products of the whole photocatalytic reaction are CO and CH_4, and their yields have reached 22.1 μmol/g and 6.0 μmol/g, respectively, after three hours of UV–vis light irradiation. In addition, the selectivity of CO can achieve 78.5%. With the increase in reaction times, the total amount of product also gradually increased, indicating that the photocatalytic CO_2 reduction was sustainable. We have reported that a new Gd(III) coordination polymer exhibited photocatalytic CO_2 reduction with a CO yield of 60.3 μmol·g^{-1} and a CO selectivity of 100% [16]. Compared with our previous results, complex (**1**) gives a different product, activity, and selectivity in photocatalytic CO_2 reduction. These results demonstrate that the environment of the catalytic center Gd(III) is vital to its catalytic activity. As a photocatalyst, the light absorption capacity is important. So, the UV–vis absorption spectrum of complex (**1**) was examined. Figure S1 (Supplementary Materials) exhibits the UV–vis absorption spectrum of the complex (**1**). It could be observed that the absorption edge of complex (**1**) is in the range of ultraviolet. Therefore, the researchers can design an idea to expand the light absorption capacity of complex (**1**) to improve its performance in photocatalytic CO_2 reduction in future studies.

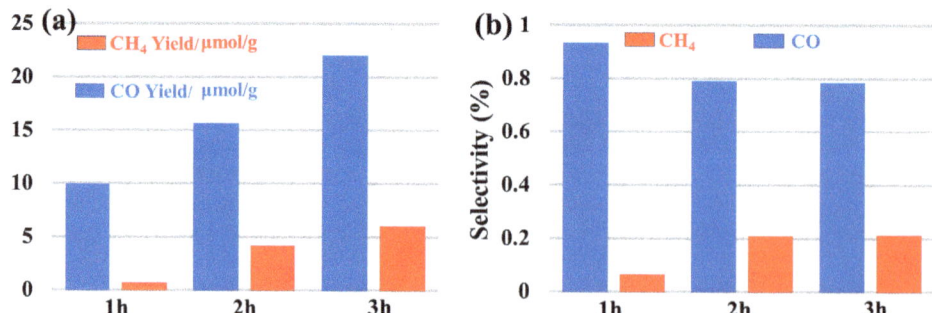

Figure 6. (**a**) Photocatalytic CO_2 reduction activity using complex (**1**); (**b**) product selectivity diagram in the photocatalytic CO_2 reduction reaction.

3. Experimental

3.1. Materials and Measurements

The materials of $GdCl_3 \cdot 6H_2O$, 6-phenylpyridine-2-carboxylic acid, 1,10-phenanthroline, and NaOH were used as received from Jilin Chinese Academy of Sciences-Yanshen Technology Co., Ltd (Changchun, China). IR spectra were recorded on a Tianjin Gangdong FTIR-850 spectrophotometer (KBr discs, range 4000~400 cm^{-1}). The Hirschfeld surface of the complex (**1**) was analyzed by the CrystalExplorer software 21.5 [25]. The crystal data of complex (**1**) were received on a Bruker CCD area detector (SuperNova, Dual, Cu at zero, 296.15 K, multi-scan).

3.2. Synthesis of Complex (1)

$GdCl_3 \cdot 6H_2O$ (0.1858 g, 0.5 mmol), 6-phenylpyridine-2-carboxylic acid (0.1992 g, 1.0 mmol), 1,10-phenanthroline (0.1802 g, 1.0 mmol), and NaOH (0.040 g, 1.0 mmol) were added to a 100 mL flask containing 30 mL of water–ethanol–DMF (*v:v:v* = 2:3:1) solution. The mixed suspension was stirred at 70 °C for 5 h and then cooled to room temperature. The colorless block crystals of complex (**1**) were obtained in four weeks.

3.3. Crystal Structure Determination

Single-crystal X-ray diffraction measurement of complex (**1**) was carried out on a Bruker CCD area detector and using Olex2 [26] for data collection at 219.98 (10) K. The structure was solved and refined with the SHELXT [17] and SHELXL [27] programs, respectively. The coordinates of hydrogen atoms were refined without any constraints or restraints. All non-hydrogen atoms were refined anisotropically. The hydrogen atoms were positioned geometrically (C–H = 0.93–0.96 Å and O–H = 0.85–1.06 Å). Their U_{iso} values were set to 1.2 U_{iso} or 1.5 U_{iso} of the parent atoms. Crystallographic data and structural refinement details of complex (**1**) are summarized in Table 4.

Table 4. Crystallographic data and structural refinement details of complex (**1**).

Empirical formula	$C_{78}H_{74}Cl_2Gd_2N_{10}O_{14}$
Formula weight	1760.87
Temperature/K	219.98(10)
Crystal system	monoclinic
Space group	*P*-1
a/Å	25.5838(6)
b/Å	13.6998(4)
c/Å	21.8202(6)
$\alpha/°$	90
$\beta/°$	90.824(3)
$\gamma/°$	90
Volume/Å^3	7647.0(4)
Z	4
ρ_{calc}, mg/mm^3	1.529
μ/mm^{-1}	1.860
S	1.060
F(000)	3544
Index ranges	$-23 \leq h \leq 30$, $-13 \leq k \leq 16$, $-25 \leq l \leq 25$
Reflections collected	17406
2θ/°	4.722–49.999
Independent reflections	6746 [R(int) = 0.0252]
Data/restraints/parameters	6746/2/484
Goodness-of-fit on F^2	1.044
Refinement method	Full-matrix least-squares on F^2
Final *R* indexes [$I \geq 2\sigma$ (*I*)]	$R_1 = 0.0328$, $wR_2 = 0.0792$
Final *R* indexes [all data]	$R_1 = 0.0397$, $wR_2 = 0.0747$

Crystallographic data for the structure reported in this paper have been deposited with the Cambridge Crystallographic Data Centre as supplementary publication No. CCDC 2292956. The CIF file can be obtained conveniently from the website: https://www.ccdc.cam.ac.uk/structures (accessed on 15 October 2023)

3.4. Photocatalytic CO_2 Reduction Evaluation

The process of photocatalytic CO_2 reduction is as follows: First, the 50 mg complex (**1**) sample was uniformly dispersed into 100 mL of deionized water H_2O in a quartz reactor and sealed. The reaction temperature was kept at 20 °C using the cooling water circulation equipment. Subsequently, high-purity CO_2 gas was bubbled into the above suspension solution with vigorous stirring for 15 min. Then, the reactor was irradiated by a 300 W Xe arc lamp (PLS-SXE300, Beijing Trusttech Co., Ltd., Beijing, China). The gas has been released every hour and tested via a gas chromatograph (FID detector, Shandong Huifen Instrument Co., Ltd., Laiwu, China).

4. Conclusions

In summary, a new dinuclear Gd(III) complex has been synthesized and characterized by IR and X-ray single-crystal diffraction analysis. The Hirschfeld surface of the complex (**1**) was analyzed. The photocatalytic CO_2 reduction experiment showed that complex (**1**) has excellent catalytic activity with yields of 22.1 µmol/g (CO) and 6.0 µmol/g (CH_4) after three hours. And the selectivity of CO can achieve 78.5%. It provides references for us to continue the study on the synthesis of rare earth metal complexes and their photocatalytic activities in the CO_2 reduction reaction.

Supplementary Materials: The following supporting information can be downloaded at: https://www.mdpi.com/article/10.3390/molecules28227595/s1. Scheme S1. Synthetic route for complex (**1**); Figure S1. UV-vis absorption spectrum of complex (**1**).

Author Contributions: L.-H.W.: conceptualization, methodology, investigation, resources, data curation, and writing—review and editing; X.-S.T.: investigation, resources, writing—review and editing, and validation. All authors have read and agreed to the published version of the manuscript.

Funding: Financial support for carrying out this work was provided by the National Natural Science Foundation of China (No. 21171132), the Science Foundation of Weifang (2020ZJ1054) and the Science Foundation of Weiyuan Scholars Innovation Team.

Institutional Review Board Statement: Not applicable.

Informed Consent Statement: Not applicable.

Data Availability Statement: Data is contained within the article and Supplementary Materials.

Conflicts of Interest: The authors declare no conflict of interest.

References

1. Liu, X.J.; Chen, T.Q.; Xue, Y.H.; Fan, J.C.; Shen, S.L.; Hossain, M.S.A.; Amin, M.A.; Pan, L.K.; Xu, X.T.; Yamauchi, Y. Nanoarchitectonics of MXene/semiconductor heterojunctions toward artificial photosynthesis via photocatalytic CO_2 reduction. *Coord. Chem. Rev.* **2022**, *459*, 214440. [CrossRef]
2. Tang, L.Q.; Jia, Y.; Zhu, Z.S.; Wu, C.P.; Zhou, Y.; Zou, Z.G. Development of functional materials for photocatalytic reduction of CO_2. *Prog. Phys.* **2021**, *41*, 254–263. [CrossRef]
3. Jiang, Z.Y.; Zhang, X.H.; Yuan, Z.M.; Chen, J.C.; Huang, B.B.; Dionysiou, D.D.; Yang, G.H. Enhanced photocatalytic CO_2 reduction via the synergistic effect between Ag and activated carbon in TiO_2/AC-Ag ternary composite. *Chem. Eng. J.* **2018**, *348*, 592–598. [CrossRef]
4. Gao, X.Q.; Cao, L.L.; Chang, Y.; Yuan, Z.Y.; Zhang, S.X.; Liu, S.J.; Zhang, M.T.; Fan, H.; Jiang, Z.Y. Improving the CO_2 Hydrogenation Activity of Photocatalysts via the Synergy between Surface Frustrated Lewis Pairs and the CuPt Alloy. *ACS Sustain. Chem. Eng.* **2023**, *11*, 5597–5607. [CrossRef]
5. Yin, H.B.; Li, J.H. New insight into photocatalytic CO_2 conversion with nearly 100% CO selectivity by CuO-Pd/H_xMoO_{3-y} hybrids. *Appl. Catal. B Environ.* **2023**, *320*, 121927. [CrossRef]
6. Heng, Q.Q.; Ma, Y.B.; Wang, X.; Wu, Y.F.; Li, Y.Z.; Chen, W. Role of Ag, Pd cocatalysts on layered $SrBi_2Ta_2O_9$ in enhancing the activity and selectivity of photocatalytic CO_2 reaction. *Appl. Surf. Sci.* **2023**, *632*, 1257564. [CrossRef]

7. Shang, X.F.; Li, G.J.; Wang, R.N.; Xie, T.; Ding, J.; Zhong, Q. Precision loading of Pd on Cu species for highly selective CO_2 photoreduction to methanol. *Chem. Eng. J.* **2023**, *456*, 140805. [CrossRef]
8. Chen, J.Y.; Li, M.; Liao, R.Z. Mechanistic insights into photochemical CO_2 reduction to CH_4 by a molecular iron-porphyrin catalyst. *Inorg. Chem.* **2023**, *62*, 9400–9417. [CrossRef]
9. Cometto, C.; Kuriki, R.; Chen, L.J.; Maeda, K.; Lau, T.C.; Ishitani, O.; Robert, M. A Carbon nitride/Fe quaterpyridine catalytic system for photostimulated CO_2-to-CO conversion with visible light. *J. Am. Chem. Soc.* **2018**, *140*, 7437–7440. [CrossRef]
10. Arikawa, Y.; Tabata, I.; Miura, Y.; Tajiri, H.; Seto, Y.; Horiuchi, S.; Sakuda, E.; Umakoshi, K. Photocatalytic CO_2 reduction under visible-light irradiation by ruthenium CNC pincer complexes. *Chem.-A Eur. J.* **2020**, *26*, 5603–5606. [CrossRef]
11. Yasuomi, Y.; Takayuki, O.; Jun, I.; Shota, F.; Chinatsu, T.; Tomoya, U.; Taro, T. Photocatalytic CO_2 reduction using various heteroleptic diimine-diphosphine Cu(I) complexes as photosensitizers. *Front. Chem.* **2019**, *7*, 288. [CrossRef]
12. Liu, D.C.; Wang, H.J.; Ouyang, T.; Wang, J.W.; Jiang, L.; Zhong, D.C.; Lu, T.B. Conjugation effect contributes to the CO_2-to-CO conversion driven by visible-light. *ACS Appl. Energy Mater.* **2018**, *1*, 2452–2459. [CrossRef]
13. Jing, H.W.; Zhao, L.; Song, G.Y.; Li, J.Y.; Wang, Z.Y.; Han, Y.; Wang, Z.X. Application of a mixed-ligand metal-organic framework in photocatalytic CO_2 reduction, antibacterial activity and dye adsorption. *Molecules* **2023**, *28*, 5204. [CrossRef]
14. Meng, S.Y.; Li, G.; Wang, P.; He, M.; Sun, X.H.; Li, Z.X. Rare earth-based MOFs for photo/electrocatalysis. *Mater. Chem. Front.* **2023**, *5*, 806–827. [CrossRef]
15. Tai, X.S.; Wang, Y.F.; Wang, L.H.; Yan, X.H. Synthesis, structural characterization, hirschfeld surface analysis and photocatalytic CO_2 reduction of Yb(III) complex with 4-aacetylphenoxyacetic acid and 1,10-phenanthroline ligands. *Bull. Chem. React. Eng. Catal.* **2023**, *18*, 285–293. [CrossRef]
16. Tai, X.S.; Wang, Y.F.; Wang, L.H.; Yan, X.H. Synthesis, structural characterization, hirschfeld surface analysis and photocatalytic CO_2 reduction activity of a new Gd(III) coordination polymer with 6-phenylpyridine-2-carboxylic acid and 4,4'-bipyridine ligands. *Bull. Chem. React. Eng. Catal.* **2023**, *18*, 353–361. [CrossRef]
17. Sheldrick, G.M. SHELXT-Integrated space-group and crystal-structure determination. *Acta Crystallogr.* **2015**, *A71*, 3–8. [CrossRef] [PubMed]
18. Li, W.J.; Chen, W.Z.; Huang, M.L. Synthesis and crystal structure of the mixed complex [Gd(Ts-p-aba)$_3$(phen)]$_2$·2DMF·4.4H$_2$O. *J. Synth. Cryst.* **2016**, *45*, 2113–2117. [CrossRef]
19. Coban, M.B. Hydrothermal synthesis, crystal structure, luminescent and magnetic properties of a new mononuclear Gd^{III} coordination complex. *J. Mol. Struct.* **2018**, *1162*, 109–116. [CrossRef]
20. Chen, L.Z.; Huang, D.D.; Cao, X.X. Synthesis, structure and dielectric properties of a novel 3D Gd(III) complex {[Gd(HPIDC)·(μ_4-C$_2$O$_4$)$_{0.5}$·H$_2$O]·2H$_2$O}$_n$. *Chin. J. Struct. Chem.* **2013**, *32*, 1553–1559. [CrossRef]
21. Becke, A.D. A new mixing of Hartree-Fock and local density-functional theories. *J. Chem. Phys.* **1993**, *98*, 1372–1377. [CrossRef]
22. Francl, M.M.; Pietro, W.J.; Hehre, W.J.; Binkley, J.S.; Gordon, M.S.; DeFrees, D.J.; Pople, J.A. Self-consistent molecular orbital methods. XXIII. A polarization-type basis set for second-row elements. *J. Chem. Phys.* **1982**, *77*, 3654–3665. [CrossRef]
23. Frisch, M.J.; Trucks, G.W.; Schlegel, H.B.; Scuseria, G.E.; Robb, M.A.; Cheeseman, J.R.; Calmani, G.; Barone, V.; Petersson, G.A.; Nakatsuji, H.; et al. *Gaussian 16, Revision C.02*; Gaussian, Inc.: Wallingford, CT, USA, 2019.
24. Lu, T.; Chen, F. Multiwfn: A multifunctional wavefunction analyzer. *J. Comput. Chem.* **2012**, *33*, 580–593. [CrossRef]
25. Spackman, P.R.; Turner, M.J.; McKinnon, J.J.; Wolff, S.K.; Grimwood, D.J.; Jayatilaka, D.; Spackman, M.A. CrystalExplorer:a program for Hirshfeld surface analysis, vis-ualization and quantitative analysis of molecular crystals. *J. Appl. Crystallogr.* **2021**, *54*, 1006–1011. [CrossRef] [PubMed]
26. Dolomanov, O.V.; Bourhis, L.J.; Gildea, R.J.; Howard, J.A.K.; Puschmann, H. OLEX2, A complete structure solution, refinement and analysis program. *J. Appl. Crystallogr.* **2009**, *42*, 339–341. [CrossRef]
27. Sheldrick, G.M. Crystal structure refinement with SHELXL. *Acta Crystallographica* **2015**, *C71*, 3–8. [CrossRef]

Disclaimer/Publisher's Note: The statements, opinions and data contained in all publications are solely those of the individual author(s) and contributor(s) and not of MDPI and/or the editor(s). MDPI and/or the editor(s) disclaim responsibility for any injury to people or property resulting from any ideas, methods, instructions or products referred to in the content.

Communication

Synthesis and Study of SrTiO₃/TiO₂ Hybrid Perovskite Nanotubes by Electrochemical Anodization

Madina Bissenova [1,2], Arman Umirzakov [1,2,3], Konstantin Mit [1], Almaz Mereke [1,3], Yerlan Yerubayev [4], Aigerim Serik [2,3] and Zhengisbek Kuspanov [2,3,*]

1. Institute of Physics and Technology, Almaty 050032, Kazakhstan; m-bisenova@list.ru (M.B.); arman_umirzakov@mail.ru (A.U.); konstantin-mit@yandex.ru (K.M.)
2. Institute of Nuclear Physics, Almaty 050032, Kazakhstan; aigerimserik3508@gmail.com
3. Department of Materials Science, Nanotechnology and Engineering Physics, Satbaev University, Almaty 050032, Kazakhstan
4. Department of Mechanics and Mechanical Engineering, M.Kh. Dulaty Taraz Regional University, Taraz 080000, Kazakhstan
* Correspondence: zhenis.kuspanov@gmail.com; Tel.: +7-707-605-04-64

Citation: Bissenova, M.; Umirzakov, A.; Mit, K.; Mereke, A.; Yerubayev, Y.; Serik, A.; Kuspanov, Z. Synthesis and Study of SrTiO₃/TiO₂ Hybrid Perovskite Nanotubes by Electrochemical Anodization. *Molecules* **2024**, *29*, 1101. https://doi.org/10.3390/molecules29051101

Academic Editors: Isabella Natali Sora and Lin Ju

Received: 24 January 2024
Revised: 23 February 2024
Accepted: 23 February 2024
Published: 29 February 2024

Copyright: © 2024 by the authors. Licensee MDPI, Basel, Switzerland. This article is an open access article distributed under the terms and conditions of the Creative Commons Attribution (CC BY) license (https://creativecommons.org/licenses/by/4.0/).

Abstract: Layers of TiO₂ nanotubes formed by the anodization process represent an area of active research in the context of innovative energy conversion and storage systems. Titanium nanotubes (TNTs) have attracted attention because of their unique properties, especially their high surface-to-volume ratio, which makes them a desirable material for various technological applications. The anodization method is widely used to produce TNTs because of its simplicity and relative cheapness; the method enables precise control over the thickness of TiO₂ nanotubes. Anodization can also be used to create decorative and colored coatings on titanium nanotubes. In this study, a combined structure including anodic TiO₂ nanotubes and SrTiO₃ particles was fabricated using chemical synthesis techniques. TiO₂ nanotubes were prepared by anodizing them in ethylene glycol containing NH₄F and H₂O while applying a voltage of 30 volts. An anode nanotube array heat-treated at 450 °C was then placed in an autoclave filled with dilute SrTiO₃ solution. Scanning electron microscopy (SEM) analysis showed that the TNTs were characterized by clear and open tube ends, with an average outer diameter of 1.01 μm and an inner diameter of 69 nm, and their length is 133 nm. The results confirm the successful formation of a structure that can be potentially applied in a variety of applications, including hydrogen production by the photocatalytic decomposition of water under sunlight.

Keywords: photocatalyst; TiO₂; TNT; SrTiO₃; anodizing

1. Introduction

Rapid growth in the world population has increased the demand for energy, the bulk of which is provided by fossil fuels for power generation, industrial needs, and transportation [1–4]. However, in addition to limited availability, the use of fossil fuels has a negative impact on the environment, creating by-products such as carbon, nitrogen, and sulfur oxides [5,6]. Therefore, there is an urgent need to develop cleaner alternative energy sources that are sustainable and have a minimal impact on the environment [7,8]. Hydrogen stands out as a clean and efficient energy source, and its production is becoming an important challenge in the field of sustainable energy sources.

Water is an abundant source of hydrogen, but given the need to introduce energy to overcome the energy barrier associated with chemical stability, it is difficult to separate water into stoichiometric hydrogen and oxygen on an industrial scale. Nevertheless, one method of hydrogen production is the photocatalytic splitting of water. Photocatalysis can efficiently utilize solar energy to split water into its individual elements [9–11]. This is a unique and promising method of hydrogen production based on the use of solar energy to convert water into hydrogen and oxygen [12]. This process could be an important step toward sustainable energy sources, as it combines the efficiency of solar panels with the ability to produce clean hydrogen. In particular, photocatalysis has been shown to be a more efficient form of wastewater treatment because of the impressive efficiency of photocatalytic removal, rapid oxidation process, lower costs, and a lack of toxicity [13]. Photocatalytic water splitting involves the use of semiconductors as photocatalysts. The most studied photocatalysts are TiO_2, ZnO, CdS, and $SrTiO_3$, which are used for various photocatalytic applications including photocatalytic water splitting [14–16]. To achieve efficient photocatalytic water splitting, a sophisticated photocatalyst is required that can overcome problems in the water oxidation process.

Titanium dioxide (titanium, TiO_2) is considered the most promising and versatile material. Over the past decades, TiO_2 has been extensively investigated in various fields because of its unique properties such as outstanding corrosion resistance, high biocompatibility, suitable bandgap for water splitting, and stable physicochemical characteristics [17,18]. The narrow–bandgap facilitates the more efficient collection of solar energy, making it an ideal material for creating an electron–hole pair. This pair is actively involved in redox reactions and finds applications in various fields, such as dyes, food, biomedicine, photocatalysis, photodegradation of water, photosensitive materials, dye-sensitized solar cells, and gas-sensitive devices. In recent years, significant research efforts have been devoted to the development of new nanomaterials, including nanostructured titanium obtained via anodization, sol–gel, hydrothermal treatment, and vapor deposition techniques [19]. Nowadays, a wide range of materials are required to develop and research advanced devices suitable for various commercial applications. Nanomaterials play a key role in emerging technologies, enabling the creation of high-performance devices [20,21]. The performance of such devices is largely determined by the geometry, shape, and morphology of the nanostructures [7]. The exponential growth in the literature indicates that interest in the nanoscale began in the 1990s. Interest in the nanoscale is driven by the commercial availability of tools used to manipulate and measure nanoscale characteristics for several reasons: (1) the anticipation of the novel physical, chemical, and biological properties of nanostructures; (2) the assumption that nanostructures will provide new building blocks for innovative materials with unique properties; (3) the miniaturization of the semiconductor industry to the nanoscale; and (4) the recognition that molecular mechanisms in biological cells function at the nanoscale [22].

2. Results and Discussion

The morphology of the obtained $SrTiO_3$ samples was studied using scanning (SEM) and transmission (TEM) electron microscopes at different resolutions. The scanning electron microscopy results (Figure 1a–c) show that the $SrTiO_3$ particles that calcined at 900 °C possess cubic shapes and have sizes ranging from 150 nm to 300 nm. Calcination at 800 °C leads to the formation of finer particles but with more significant numbers of impurities such as $SrCO_3$. Based on the literature and experimental data [23,24], the optimal calcination temperature is 900 °C, which is followed by treatment in 1 m nitric acid solution to remove residual $SrCO_3$. However, it is worth noting that the particle sizes are highly heterogeneous. Given studies in the literature, doping $SrTiO_3$ with other elements, such as Al or Mn, can contribute to the size reduction and distortion in $SrTiO_3$'s crystal shape.

In the case of TEM, clearly formed cubes of $SrTiO_3$ with anisotropic structures with an average size of about 200 nm are clearly visible, as shown in Figure 1d–f. An important feature of these particles is the anisotropic structure, which creates a difference in energy at different faces, leading to the formation of p–n junctions. This allows the charge within each photocatalyst particle to be separated using an inter-domain electric field. Thus, electrons are concentrated on some faces and holes on other faces, which provides for the separation of photocatalytic reduction and oxidation processes on different faces. Given the anisotropic crystal structure, the selective deposition of catalysts takes place, which leads to the release of hydrogen and oxygen on the faces of the cubic photocatalyst.

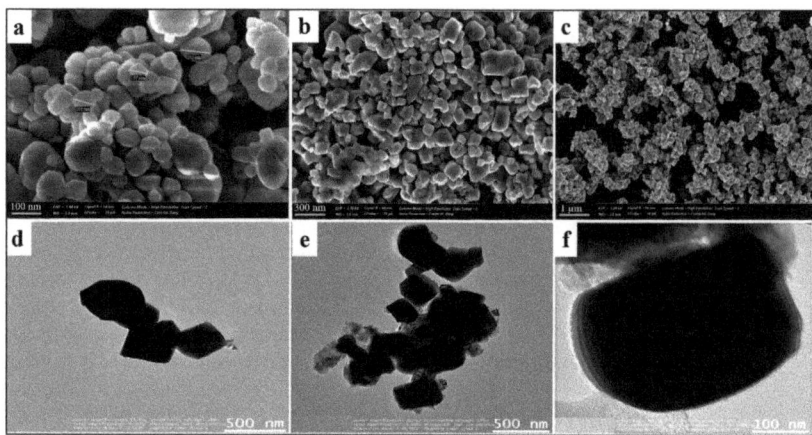

Figure 1. SEM (**a–c**) and TEM (**d–f**) images at different magnifications of cubic SrTiO$_3$ obtained by a chemical precipitation method.

Among various nanostructured oxide materials, TiO$_2$ nanotubes have been emphasized because of their improved properties, economical design, and higher surface-to-volume ratio [25]. TNTs with high specific surface areas, ion exchange abilities, and photocatalytic properties have been considered for various potential applications and can be excellent candidates as catalysts in photocatalysis [26]. Figure 2 shows images of the top surfaces of the anodized TiO$_2$ nanotube samples before and after the deposition of SrTiO$_3$ on their surfaces. The top surfaces of the anodized and annealed TNTs at 450 °C shown in Figure 2a show well-defined tubes with open ends that form a hexagonal order. This is typical of anodization, as previously noted in [27]. After applying SrTiO$_3$, pronounced morphological changes are observed with the presence of interface regions between SrTiO$_3$ and TNT, which indicates the success of the combination (Figure 2b). During 6 h of autoclave treatment, the surface showed a tendency to be coated with nanoparticles, and uneven deposition was also found. Agglomerates are formed on the surface, and round holes corresponding to TNT are still visible. Note that increasing the treatment time to 6 h significantly affects the surface morphology, leading to the formation of larger agglomerates and the blocking of the tube tops [28]. Figure 2c shows that the initial TNTs have an average outer diameter of 1 µm, while Figure 2a shows an inner diameter of 69 nm. The length of the tubes is 133 nm, as seen in the inset. A cross-sectional view of a freestanding titanium dioxide membrane with an average thickness of more than 50 nm is shown in Figure 2d, mechanically collapsed for visualization. In a related study [25], Paulose et al. obtained nanotubes measuring 360 µm in length over a 96 h period, utilizing a voltage of 60 V. They employed a titanium foil with a thickness of 0.25 mm, immersed in a solution comprising 0.3 wt% NH$_4$F and 2% H$_2$O in ethylene glycol. Our results—derived from anodization in a solution comprising 0.7 wt% NH$_4$F and 3.5 wt% distilled water at 30 V—revealed the length of the TNT nanotubes to be 133 nm at the nanoscale. The SEM images also demonstrate that the obtained nanotubes are ordered and have clear open ends. Despite the low voltage (30 V), we compensated for this by increasing the concentrations of NH$_4$F and H$_2$O in the anode solution. Given the higher mass percentage of NH$_4$F, compensation is accomplished by increasing the concentration of H$_2$O, resulting in faster growth and, hence, longer nanotube lengths.

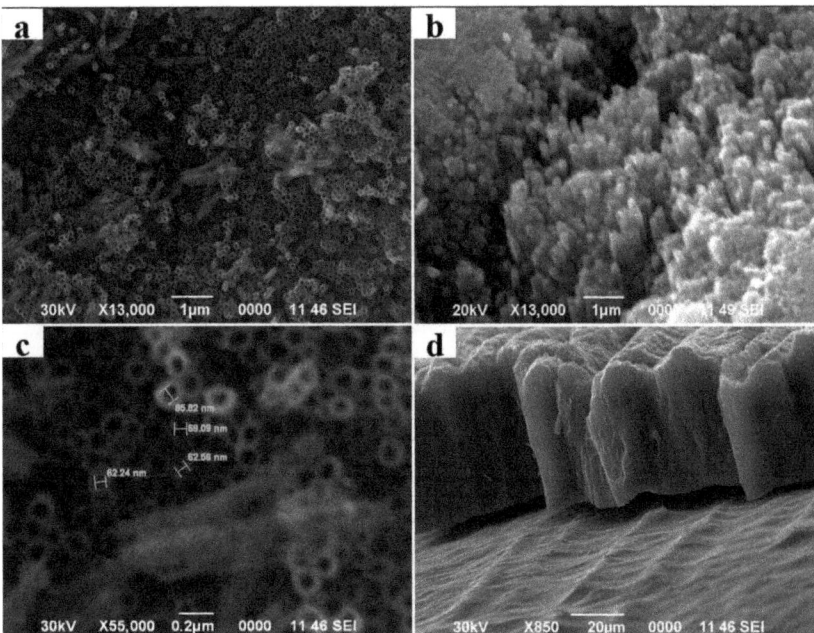

Figure 2. (a) SEM images showing the top surface of the TNT anode array; (b) the TNT@SrTiO$_3$ array; (c) the top view and (d) side view of samples with magnification of the surface of the anode array prepared at 30 V.

XRD analysis of the TNT@SrTiO$_3$ samples was performed on an X-ray diffractometer with detection unit rotation angles ranging from 20° to 80° and a minimum detection unit movement step of 0.01, as shown in Figure 3a. The characteristic peaks of the TNT samples appear at 2θ 25.4°, 37.9°, and 53.4°, 71.5°, indicating the polycrystalline structure of the anatase, in good agreement with the standard map for TNT (JCPDS map 1286) [20]. In addition, the appearance of new peaks at 32.2°, 46.9°, and 57.8° in the X-ray diffraction spectrum of the TNT@SrTiO$_3$ samples indicates the combination of two components in the composite, which additionally proves the successful connection and interaction between the components. This confirmation is based on a comparison of the diffraction spectra of the composite with TNT, which makes it possible to determine whether changes have occurred in the crystal structure during their combination. It is particularly important to note that the peak at 2θ, equal to 71.23°, has a high intensity, indicating the high crystallinity of the semiconductor. This is significant because the transport efficiency of charged carriers generated during photogeneration can be strongly dependent on the crystallinity of the material. Low crystallinity can lead to the inefficient migration of charged particles. In addition, semi-quantitative elemental analysis of the particles confirmed the composition of the obtained samples. The presence of the elements Ti and Sr was confirmed without detecting other impurities. According to the atomic percentages in the index (Figure 3b), it can be established that Ti/Sr is 81.10%/18.90%, respectively. These results confirm that the designs contain the expected elements and have no significant impurities.

Low-temperature electron paramagnetic resonance (EPR) spectra were determined on the SrTiO$_3$/TiO$_2$ samples to confirm the presence of Ti^{3+} and oxygen vacancies. The initial SrTiO$_3$/TiO$_2$ (Figure 4c, marked in red), containing mainly Ti^{4+} 3d0 states, exhibits a weak EPR signal, which may be due to the surface adsorption of O$_2$ from air. For SrTiO$_3$ and TiO$_2$, a strong signal from Ti^{3+} spins (marked in blue and black) is also observed. It is generally believed that photoelectrons can be captured by Ti^{4+} and lead to the reduction of

Ti^{4+} cations to the Ti^{3+} state, which is usually accompanied by the loss of oxygen from the surface of TiO$_2$ and SrTiO$_3$. Thus, these data clearly confirm that Ti^{3+} and oxygen vacancies were formed in all SrTiO$_3$/TiO$_2$, TiO$_2$, and SrTiO$_3$ samples.

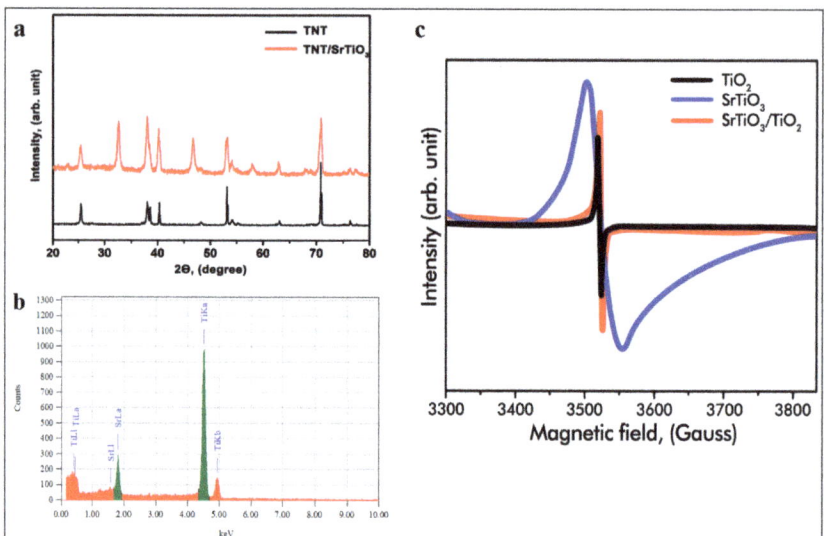

Figure 3. (a) X-ray diffraction analysis of combined TNT@ SrTiO$_3$; (b) semi-quantitative elemental analysis of TNT@SrTiO$_3$ particles; (c) EPR spectra of pristine TiO$_2$ and SrTiO$_3$ and pristine SrTiO$_3$/TiO$_2$ nanotube arrays after hydrothermal reaction of 5 h duration.

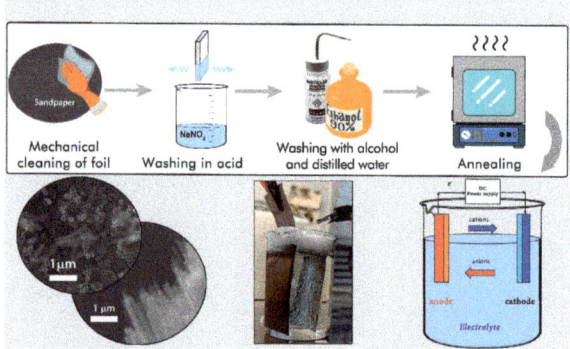

Figure 4. Schematic illustration of the stages of obtaining TNT.

TiO$_2$ nanotubes can be produced in various ways [29], among which, the most widely studied is the use of electrochemical anodization. The advantage of anodic TiO$_2$ nanotubes over TiO$_2$ nanotubes produced by other methods is their availability and cost-effectiveness. Also, one of the advantages of this method is that the anodic TiO$_2$ nanotubes grow vertically on the Ti substrate with nanotube holes on top and closed nanotube bottoms attached to the Ti substrate. Thus, no further immobilization on the substrate is required. The TNT layers are highly ordered, which favors a direct diffusion pathway. In addition, the nanotube layers can be removed from the Ti substrate and used as powders if required. Another advantage is that the nanotube layer thickness and nanotube diameter can be controlled by adjusting the anodization electrolyte, potential, and time [30].

3. Materials and Methods

3.1. Materials

Ti foil (99.9%; thickness, 0.1 mm; China), ethanol (45%), ethylene glycol (99.9%, Russia), ammonium fluoride, and sodium nitrate (70%) were used without further purification. Distilled water was used as a solvent in all experiments.

3.2. Synthesis of $SrTiO_3$

$SrTiO_3$ was obtained using a chemical precipitation method [24,31–33]. For this purpose, 2.54 g of $Sr(NO_3)_2$ was mixed with 100 mL of distilled water; then, 0.958 g of TiO_2 was added in a 1:1 ratio of Ti and $SrTiO_3$ to this solution. The solution was then treated for 30 min in an ultrasonic bath. The solution was gradually added while maintaining vigorous stirring, and the pH of the mixture was brought to 6–7 using 10% NH_3OH solution. The suspension was washed several times with distilled water. The resulting powder was dried at 60 °C overnight and then calcined at 900 °C for 1 h.

3.3. Nanotube Synthesis

TiO_2 was obtained using an anodization method. The 0.1 mm thick Ti foil was initially cut into 1 cm × 6 cm samples and mechanically polished with P150 sandpaper. The sheets were then ultrasonically treated in sodium nitrate, ethanol, and distilled water for final cleaning. Electrochemical anodization experiments were carried out in a two-electrode electrochemical cell, where titanium foil served as the working electrode and a sheet of nickel foil as the counter electrode at constant potential and room temperature (≈22 °C). Figure 1 shows a schematic of the titanium nanotube formation process. A constant current power supply unit model, UNI-T UTP3315TPL from UNI-TREND Technology, China, was used. This unit was used as a voltage source to control the anodization. The electrolyte for anodizing consisted of ethylene glycol with 0.7 wt% NH_4F and 3.5 wt% distilled water added. The anodization process was carried out at 30 V for 96 h at room temperature. The anodized titanium nanotube samples were then placed in ethylene glycol and subjected to ultrasonic stirring until the nanotube film separated from the titanium substrate. The suspension was filtered; the residue was washed several times with distilled water. The resulting powder was dried at 60 °C for 3 h and then calcined at 450 °C for 1 h.

3.4. Synthesis of $TNT@SrTiO_3$

To create the combined $TNT@SrTiO_3$ structure, powders of 0.2 g of TNT and 0.1 g of $SrTiO_3$ were taken, mixed with 40 mL of distilled water, and placed in a stainless autoclave. The sealed autoclave was heated to 90 °C and incubated for 6 h. At the end of the experiment, the autoclave was cooled to room temperature. The samples were then washed with distilled water and dried in an oven for 5 h at 60 °C.

3.5. Material Characterization Techniques

The morphologies of the TNT and the combined $TNT@SrTiO_3$ were analyzed using a JSM-6490LA scanning electron microscope from JEOL, Tokyo, Japan. TESCAN MAIA3 XMU scanning transmission electron microscopy (STEM) was used to further investigate the morphology at high resolution. The crystal structure of the samples was studied using a Drone-8 X-ray diffractometer. An EPR spectrometer "JEOL" (JES-FA200, Japan) was also used. Measurements were in ranges of ~9.4 GHz (X-Band) and ~35 GHz (Q-Band). Microwave frequency stability—~10^{-6}. Sensitivity—$7 \times 10^9/10^{-4}$ Tl. Resolution—2.35 µT. Output power—from 200 mW to 0.1 µW. Quality factor (Q-factor)—18,000.

4. Conclusions

In this paper, the synthesis of arrays of TiO_2 nanotubes using an electrochemical anodization method was successfully demonstrated. The obtained nanotubes have clear and open ends and are 133.9 nm long, and their membranes are more than 1 µm thick. The anodization process of 0.1 mm thick Ti foil at 450 °C can easily produce such nanotube arrays. SEM analysis showed that the TNTs are characterized by clear and open tube ends, with an average outer diameter of 1 μm and an inner diameter of 69 nm, and their length is 133 nm. In addition, a combined structure of TNT@SrTiO$_3$ was fabricated in this study using chemical autoclave synthesis techniques. X-ray phase analysis confirmed the high crystallinity and orientation of crystallites along the preferential growth direction, indicating the successful formation of the structure. The results obtained here have potential significance for various fields including the sunlight-induced photocatalytic decomposition of water and other applications in energy conversion and storage. Further research and development in this area can contribute to the development of innovative technologies and improve the efficiency of energy systems.

Author Contributions: Conceptualization, methodology, writing—original draft preparation, investigation: M.B.; resources, visualization, project administration, funding acquisition: A.U., K.M., A.M. and Y.Y.; formal analysis: A.S.; supervision, data curation, writing—review and editing: Z.K. All authors have read and agreed to the published version of the manuscript.

Funding: This research was funded by the Science Committee of the Ministry of Science and Higher Education and of the Republic of Kazakhstan (Grant No AP19680604).

Institutional Review Board Statement: Not applicable.

Informed Consent Statement: Not applicable.

Data Availability Statement: The data are contained within the article.

Conflicts of Interest: There are no conflicts of interest to declare.

References

1. Zhang, Z.; Wang, Q.; Xu, H.; Zhang, W.; Zhou, Q.; Zeng, H.; Yang, J.; Zhu, J.; Zhu, X. TiO$_2$ Nanotube Arrays with a Volume Expansion Factor Greater than 2.0: Evidence against the Field-Assisted Ejection Theory. *Electrochem. Commun.* **2020**, *114*, 106717. [CrossRef]
2. Prikhodko, N.; Yeleuov, M.; Abdisattar, A.; Askaruly, K.; Taurbekov, A.; Tolynbekov, A.; Rakhymzhan, N.; Daulbayev, C. Enhancing Supercapacitor Performance through Graphene Flame Synthesis on Nickel Current Collectors and Active Carbon Material from Plant Biomass. *J. Energy Storage* **2023**, *73*, 108853. [CrossRef]
3. Askaruly, K.; Yeleuov, M.; Taurbekov, A.; Sarsembayeva, B.; Tolynbekov, A.; Zhylybayeva, N.; Azat, S.; Abdisattar, A.; Daulbayev, C. A Facile Synthesis of Graphite-Coated Amorphous SiO$_2$ from Biosources as Anode Material for Libs. *Mater. Today Commun.* **2023**, *34*, 105136. [CrossRef]
4. Yeleuov, M.; Daulbayev, C.; Taurbekov, A.; Abdisattar, A.; Ebrahim, R.; Kumekov, S.; Prikhodko, N.; Lesbayev, B.; Batyrzhan, K. Synthesis of Graphene-like Porous Carbon from Biomass for Electrochemical Energy Storage Applications. *Diam. Relat. Mater.* **2021**, *119*, 108560. [CrossRef]
5. Galstyan, V.; Macak, J.M.; Djenizian, T. Anodic TiO$_2$ Nanotubes: A Promising Material for Energy Conversion and Storage. *Appl. Mater. Today* **2022**, *29*, 101613. [CrossRef]
6. Taurbekov, A.; Abdisattar, A.; Atamanov, M.; Yeleuov, M.; Daulbayev, C.; Askaruly, K.; Kaidar, B.; Mansurov, Z.; Castro-Gutierrez, J.; Celzard, A.; et al. Biomass Derived High Porous Carbon via CO$_2$ Activation for Supercapacitor Electrodes. *J. Compos. Sci.* **2023**, *7*, 444. [CrossRef]
7. Saddique, Z.; Imran, M.; Javaid, A.; Kanwal, F.; Latif, S.; dos Santos, J.C.S.; Kim, T.H.; Boczkaj, G. Bismuth-Based Nanomaterials-Assisted Photocatalytic Water Splitting for Sustainable Hydrogen Production. *Int. J. Hydrogen Energy* **2024**, *52*, 594–611. [CrossRef]
8. Yergaziyeva, G.; Kuspanov, Z.; Mambetova, M.; Khudaibergenov, N.; Makayeva, N.; Daulbayev, C. Advancements in Catalytic, Photocatalytic, and Electrocatalytic CO$_2$ Conversion Processes: Current Trends and Future Outlook. *J. CO2 Util.* **2024**, *80*, 102682. [CrossRef]
9. Wang, S.-J.; Su, D.; Zhu, Y.-F.; Lu, C.-H.; Zhang, T. The State-of-the-Art Review on Rational Design for Cavitation Assisted Photocatalysis. *Mater. Des.* **2023**, *234*, 112377. [CrossRef]
10. Kuspanov, Z.; Bakbolat, B.; Baimenov, A.; Issadykov, A.; Yeleuov, M.; Daulbayev, C. Photocatalysts for a Sustainable Future: Innovations in Large-Scale Environmental and Energy Applications. *Sci. Total Environ.* **2023**, *885*, 163914. [CrossRef]
11. Megbenu, H.K.; Daulbayev, C.; Nursharip, A.; Tauanov, Z.; Poulopoulos, S.; Busquets, R.; Baimenov, A. Photocatalytic and Adsorption Performance of MXene@Ag/Cryogel Composites for Sulfamethoxazole and Mercury Removal from Water Matrices. *Environ. Technol. Innov.* **2023**, *32*, 103350. [CrossRef]
12. Baimenov, A.; Montagnaro, F.; Inglezakis, V.J.; Balsamo, M. Experimental and Modeling Studies of Sr^{2+} and Cs$^+$ Sorption on Cryogels and Comparison to Commercial Adsorbents. *Ind. Eng. Chem. Res.* **2022**, *61*, 8204–8219. [CrossRef]

13. Lee, D.-E.; Kim, M.-K.; Danish, M.; Jo, W.-K. State-of-the-Art Review on Photocatalysis for Efficient Wastewater Treatment: Attractive Approach in Photocatalyst Design and Parameters Affecting the Photocatalytic Degradation. *Catal. Commun.* **2023**, *183*, 106764. [CrossRef]
14. Indira, K.; Mudali, U.K.; Nishimura, T.; Rajendran, N. A Review on TiO_2 Nanotubes: Influence of Anodization Parameters, Formation Mechanism, Properties, Corrosion Behavior, and Biomedical Applications. *J. Bio- Tribo-Corros.* **2015**, *1*, 28. [CrossRef]
15. Kuspanov, Z.; Umirzakov, A.; Serik, A.; Baimenov, A.; Yeleuov, M.; Daulbayev, C. Multifunctional Strontium Titanate Perovskite-Based Composite Photocatalysts for Energy Conversion and Other Applications. *Int. J. Hydrogen Energy* **2023**, *48*, 38634–38654. [CrossRef]
16. Bakbolat, B.; Daulbayev, C.; Sultanov, F.; Beissenov, R.; Umirzakov, A.; Mereke, A.; Bekbaev, A.; Chuprakov, I. Recent Developments of TiO_2-Based Photocatalysis in the Hydrogen Evolution and Photodegradation: A Review. *Nanomaterials* **2020**, *10*, 1790. [CrossRef] [PubMed]
17. Moon, S.; Nagappagari, L.R.; Lee, J.; Lee, H.; Lee, W.; Lee, K. Electrochemical Detection of 2,4,6-Trinitrotoluene Reduction in Aqueous Solution by Using Highly Ordered 1D TiO_2 Nanotube Arrays. *Mater. Today Commun.* **2020**, *25*, 101389. [CrossRef]
18. An, X.; Hua, W.; Rui, L.; Liu, P.; Li, X. Application of a New Nano-TiO_2 Composite Antibacterial Agent in Nursing Management of Operating Room: Based on Real-Time Information Push Assistant System. *Prev. Med.* **2023**, *172*, 107541. [CrossRef]
19. Yin, H.; Liu, H.; Shen, W.Z. The Large Diameter and Fast Growth of Self-Organized TiO_2 Nanotube Arrays Achieved via Electrochemical Anodization. *Nanotechnology* **2009**, *21*, 035601. [CrossRef]
20. Ghani, T.; Mujahid, M.; Mehmood, M.; Zhang, G.; Naz, S. Highly Ordered Combined Structure of Anodic TiO_2 Nanotubes and TiO_2 Nanoparticles Prepared by a Novel Route for Dye-Sensitized Solar Cells. *J. Saudi Chem. Soc.* **2019**, *23*, 1231–1240. [CrossRef]
21. Jandosov, J.; Alavijeh, M.; Sultakhan, S.; Baimenov, A.; Bernardo, M.; Sakipova, Z.; Azat, S.; Lyubchyk, S.; Zhylybayeva, N.; Naurzbayeva, G.; et al. Activated Carbon/Pectin Composite Enterosorbent for Human Protection from Intoxication with Xenobiotics Pb(II) and Sodium Diclofenac. *Molecules* **2022**, *27*, 2296. [CrossRef]
22. Zhang, W.; Sun, Y.; Tian, R.; Gao, Q.; Wang, Y.; Liu, Y.; Yang, F. Anodic Growth of TiO_2 Nanotube Arrays: Effects of Substrate Curvature and Residual Stress. *Surf. Coat. Technol.* **2023**, *469*, 129783. [CrossRef]
23. Roy, P.K.; Bera, J. Formation of $SrTiO_3$ from Sr-Oxalate and TiO_2. *Mater. Res. Bull.* **2005**, *40*, 599–604. [CrossRef]
24. Kudaibergen, A.D.; Kuspanov, Z.B.; Issadykov, A.N.; Beisenov, R.E.; Mansurov, Z.A.; Yeleuov, M.A.; Daulbayev, C.B. Synthesis, Structure, and Energetic Characteristics of Perovskite Photocatalyst SrTiO3: An Experimental and DFT Study. *Eurasian Chem.-Technol. J.* **2023**, *25*, 139–146. [CrossRef]
25. Paulose, M.; Prakasam, H.E.; Varghese, O.K.; Peng, L.; Popat, K.C.; Mor, G.K.; Desai, T.A.; Grimes, C.A. TiO_2 Nanotube Arrays of 1000 μm Length by Anodization of Titanium Foil: Phenol Red Diffusion. *J. Phys. Chem. C* **2007**, *111*, 14992–14997. [CrossRef]
26. Moulis, F.; Krýsa, J. Photocatalytic Degradation of Several VOCs (*n*-Hexane, *n*-Butyl Acetate and Toluene) on TiO_2 Layer in a Closed-Loop Reactor. *Catal. Today* **2013**, *209*, 153–158. [CrossRef]
27. Hou, X.; Lund, P.D.; Li, Y. Controlling Anodization Time to Monitor Film Thickness, Phase Composition and Crystal Orientation during Anodic Growth of TiO_2 Nanotubes. *Electrochem. Commun.* **2022**, *134*, 107168. [CrossRef]
28. Sopha, H.; Samoril, T.; Palesch, E.; Hromadko, L.; Zazpe, R.; Skoda, D.; Urbanek, M.; Ng, S.; Prikryl, J.; Macak, J.M. Ideally Hexagonally Ordered TiO_2 Nanotube Arrays. *ChemistryOpen* **2017**, *6*, 480–483. [CrossRef] [PubMed]
29. Beketova, D.; Motola, M.; Sopha, H.; Michalicka, J.; Cicmancova, V.; Dvorak, F.; Hromadko, L.; Frumarova, B.; Stoica, M.; Macak, J.M. One-Step Decoration of TiO_2 Nanotubes with Fe_3O_4 Nanoparticles: Synthesis and Photocatalytic and Magnetic Properties. *ACS Appl. Nano Mater.* **2020**, *3*, 1553–1563. [CrossRef]
30. Tak, M.; Tomar, N.; Mote, R.G. Synthesis of Titanium Nanotubes (TNT) and Its Influence on Electrochemical Micromachining of Titanium. *Procedia CIRP* **2020**, *95*, 803–808. [CrossRef]
31. Daulbayev, C.; Sultanov, F.; Korobeinyk, A.V.; Yeleuov, M.; Azat, S.; Bakbolat, B.; Umirzakov, A.; Mansurov, Z. Bio-Waste-Derived Few-Layered Graphene/$SrTiO_3$/PAN as Efficient Photocatalytic System for Water Splitting. *Appl. Surf. Sci.* **2021**, *549*, 149176. [CrossRef]
32. Sultanov, F.; Daulbayev, C.; Azat, S.; Kuterbekov, K.; Bekmyrza, K.; Bakbolat, B.; Bigaj, M.; Mansurov, Z. Influence of Metal Oxide Particles on Bandgap of 1D Photocatalysts Based on $SrTiO_3$/PAN Fibers. *Nanomaterials* **2020**, *10*, 1734. [CrossRef] [PubMed]
33. Sultanov, F.; Daulbayev, C.; Bakbolat, B.; Daulbayev, O.; Bigaj, M.; Mansurov, Z.; Kuterbekov, K.; Bekmyrza, K. Aligned Composite $SrTiO_3$/PAN Fibers as 1D Photocatalyst Obtained by Electrospinning Method. *Chem. Phys. Lett.* **2019**, *737*, 136821. [CrossRef]

Disclaimer/Publisher's Note: The statements, opinions and data contained in all publications are solely those of the individual author(s) and contributor(s) and not of MDPI and/or the editor(s). MDPI and/or the editor(s) disclaim responsibility for any injury to people or property resulting from any ideas, methods, instructions or products referred to in the content.

Article

Optical and Photocatalytic Properties of Cobalt-Doped LuFeO$_3$ Powders Prepared by Oxalic Acid Assistance

Zhi Wang [1,2,*], Changmin Shi [2], Pengfei Li [1], Wenzhu Wang [1], Wenzhen Xiao [2], Ting Sun [1] and Jing Zhang [1,*]

1. School of Physics and Electrical Engineering, Anyang Normal University, Anyang 455000, China
2. School of Physics and Electronic Engineering, Linyi University, Linyi 276000, China
* Correspondence: wangzhi@lyu.edu.cn (Z.W.); zj@aynu.edu.cn (J.Z.); Tel.: +86-178-0613-0619 (Z.W.)

Abstract: B-site cobalt (Co)-doped rare-earth orthoferrites ReFeO$_3$ have shown considerable enhancement in physical properties compared to their parent counterparts, and Co-doped LuFeO$_3$ has rarely been reported. In this work, LuFe$_{1-x}$Co$_x$O$_3$ (x = 0, 0.05, 0.1, 0.15) powders have been successfully prepared by a mechanochemical activation-assisted solid-state reaction (MAS) method at 1100 °C for 2 h. X-ray diffraction (XRD) and Fourier transform infrared (FTIR) spectroscopy studies demonstrated that a shrinkage in lattice parameters emerges when B-site Fe ions are substituted by Co ions. The morphology and elemental distribution were investigated by scanning electron microscopy (SEM) and energy dispersive spectroscopy (EDS). The UV–visible absorbance spectra show that LuFe$_{0.85}$Co$_{0.15}$O$_3$ powders have a narrower bandgap (1.75 eV) and higher absorbance than those of LuFeO$_3$ (2.06 eV), obviously improving the light utilization efficiency. Additionally, LuFe$_{0.85}$Co$_{0.15}$O$_3$ powders represent a higher photocatalytic capacity than LuFeO$_3$ powders and can almost completely degrade MO in 5.5 h with the assistance of oxalic acid under visible irradiation. We believe that the present study will promote the application of orthorhombic LuFeO$_3$ in photocatalysis.

Keywords: cobalt doping; rare-earth orthoferrites; optical properties; photocatalysis; hole scavengers

Citation: Wang, Z.; Shi, C.; Li, P.; Wang, W.; Xiao, W.; Sun, T.; Zhang, J. Optical and Photocatalytic Properties of Cobalt-Doped LuFeO$_3$ Powders Prepared by Oxalic Acid Assistance. *Molecules* **2023**, *28*, 5730. https://doi.org/10.3390/molecules28155730

Academic Editor: Igor Reva

Received: 28 June 2023
Revised: 17 July 2023
Accepted: 23 July 2023
Published: 28 July 2023

Copyright: © 2023 by the authors. Licensee MDPI, Basel, Switzerland. This article is an open access article distributed under the terms and conditions of the Creative Commons Attribution (CC BY) license (https://creativecommons.org/licenses/by/4.0/).

1. Introduction

In recent decades, a flurry of research has been devoted to studying perovskite-type rare-earth ferrites, RFeO$_3$ (R denotes rare-earth), due to their potential applications in data storage devices [1,2], catalysis [3,4], solid-oxide cells [5], gas sensors [6], etc. Since the ABO$_3$ perovskite structure has a large tolerance for structural distortion [7], the electrical, optical, and magnetic properties can be effectively modified via tuning spin, charge, orbital, and lattice coupling by substituting A- and/or B-sites to meet the desired demands in applications, which makes RFeO$_3$ compounds rather appealing [8].

Among these applications, photocatalysis is generally referred to as the catalysis of a photochemical reaction at a solid surface, usually a semiconductor. For light-harvesting and visible light-driven photocatalysis, the photocatalytic capacity can be improved by bandgap engineering via doping [9], providing more active sites via the design of active nanostructures [5], preventing electron–hole recombination via involving electron–hole scavengers [3], enhancing adsorption via introducing vacancies [10,11], etc. Bandgap engineering via doping is challenging in these approaches. Firstly, the preparation method should be effective in replacing the atoms at the target sites. Secondly, semiconductor doping can simultaneously alter its bandgap and electronegativity, which determine the redox potential at the conduction band minimum (CBM) and the valence band maximum (VBM) [12], and the redox potential plays a key role in the photocatalysis process.

LuFeO$_3$, as an important member of the ReFeO$_3$ family, is valued for its stability, high Neel temperature, large magnetocrystalline anisotropy, and Dzyaloshinskii–Moriya interaction [13–15]. More recently, Zhou et al. [3] reported orthorhombic LuFeO$_3$ as a new photocatalyst for dye degradation. They pointed out that the reaction of holes with

OH$^-$ is possibly not the dominant way to generate the free radical ·OH. This increases the possibility of improving catalytic efficiency via doping because the effect of VBM oxidation potential can be weakened to some extent.

Metal doping is a well-known strategy to enhance the electrical, chemical, and optical performance of materials, such as the durability of Li-ion batteries [16], the pseudocapacitive activities of WS$_2$ [17], the nonlinear optical response of aluminum nitride nanocages [18], the hydrogen evolution reduction of biomass-based carbon materials [19], the electrochemical activities of MnO$_2$ [20], etc. Among these metals, cobalt (Co) is considered an interesting dopant because of its high Pauling electronegativity number (1.88) [21], various oxidation levels, and spin states [22,23]. Recently, researchers have carried out extensive investigations on the structure, morphology, and electrical, optical, and magnetic properties of B-site cobalt-doped orthorhombic RFeO$_3$. Most research focuses on orthorhombic NdFeO$_3$ [24–28], YbFeO$_3$ [29], DyFeO$_3$ [30], and LaFeO$_3$ [31]. It has been found that these Co-doped RFeO$_3$ systems show considerable enhancement in their physical properties compared to their parent counterparts. However, as orthorhombic LuFeO$_3$ has the smallest tolerance factor (0.866) [32], its perovskite structure is the most unstable in RFeO$_3$, so it is difficult to prepare B-site Co-doped LuFeO$_3$, which has rarely been reported. In addition, there are few reports on the elimination of impurity phases in the preparation of Co-doped RFeO$_3$, although it is important to eliminate the influence of impurities on the properties of pure-phase ReFeO$_3$ systems.

The powders prepared by the solid-state reaction method, which usually results in large particle sizes and poor morphology, are seldomly employed to perform photocatalytic experiments. However, the mechanochemical activation-assisted solid-state reaction (MAS) method, a modified solid-state reaction method reported in our previous work [33], enables the preparation of desired powders at a relatively low temperature for a relatively short time. In this paper, we prepared B-site Co-doped LuFeO$_3$ powders by the MAS method reported in our previous work [33]. The preparation, fine structure, morphology, and optical properties are investigated by X-ray diffraction (XRD), Fourier transform infrared spectroscopy (FTIR), scanning electron microscopy (SEM), energy dispersive spectroscopy (EDS), and UV–visible absorbance spectroscopy. The photocatalytic capacity was investigated by degrading methyl orange (MO) assisted by oxalic acid (OA) under visible light irradiation. A possible degradation mechanism is proposed.

2. Results and Discussion

2.1. Preparation of Co-Doped LuFeO$_3$ Powders and Structure Analysis

The effects of Co doping amount, calcination time, and temperature on lutetium ferrite phase evolution were studied by XRD, and the optimum preparation parameters were obtained. The results are represented in Figure 1. As shown in Figure 1a, the LuFe$_{0.9}$Co$_{0.1}$O$_3$ sample (marked in red) was attempted to be prepared using the same calcination as our previous work, at 1200 °C for 10 h [33], but a distinct peak at 29.8° corresponding to the impurity Lu$_2$O$_3$ was observed, as marked with a red ellipse. When the calcination time was reduced to 2 h (marked in blue), the content of Lu$_2$O$_3$ decreased significantly, which indicates that Co doping enables the system to be prepared at a lower energy level. Thus, the effects of the calcination temperature on the phase evolution of the lutetium ferrite system were studied by increasing the reaction temperature from 1050 °C to 1300 °C, as shown in Figure 1b. It can be seen that a low or high calcination temperature is not beneficial to obtaining the target product. The optimum preparation parameters for the LuFe$_{0.9}$Co$_{0.1}$O$_3$ sample are calcination at 1100 °C for 2 h. Under this condition, a series of LuFe$_{1-x}$Co$_x$O$_3$ (x = 0, 0.1, 0.15, 0.2, 0.25) samples were attempted to be synthesized, and the results are shown in Figure 1c. It can be observed that the peak at 29.8°, corresponding to Lu$_2$O$_3$, appears when x reaches 0.2. As mentioned in the Introduction, because the perovskite structure of orthorhombic LuFeO$_3$ is most unstable in ReFeO$_3$ systems, the Co doping level should be lower than in other RFeO$_3$ systems such as NdFeO$_3$ [28] and DyFeO$_3$ [30]. Further, the XRD patterns in Figure 1c indicate that the 2θ angle values of the (020), (112), and (200) diffraction peaks gradually shift to a higher angle after Co was included in the LuFeO$_3$ lattice when no Lu$_2$O$_3$ peak emerged in the samples. The highlighted (112) plane shows an obvious change in the peak position. The d-spacings of the Co-doped LuFeO$_3$ lattice should be narrowed by Co doping, according to Bragg's law. Thus, a shrinkage in lattice parameters is expected.

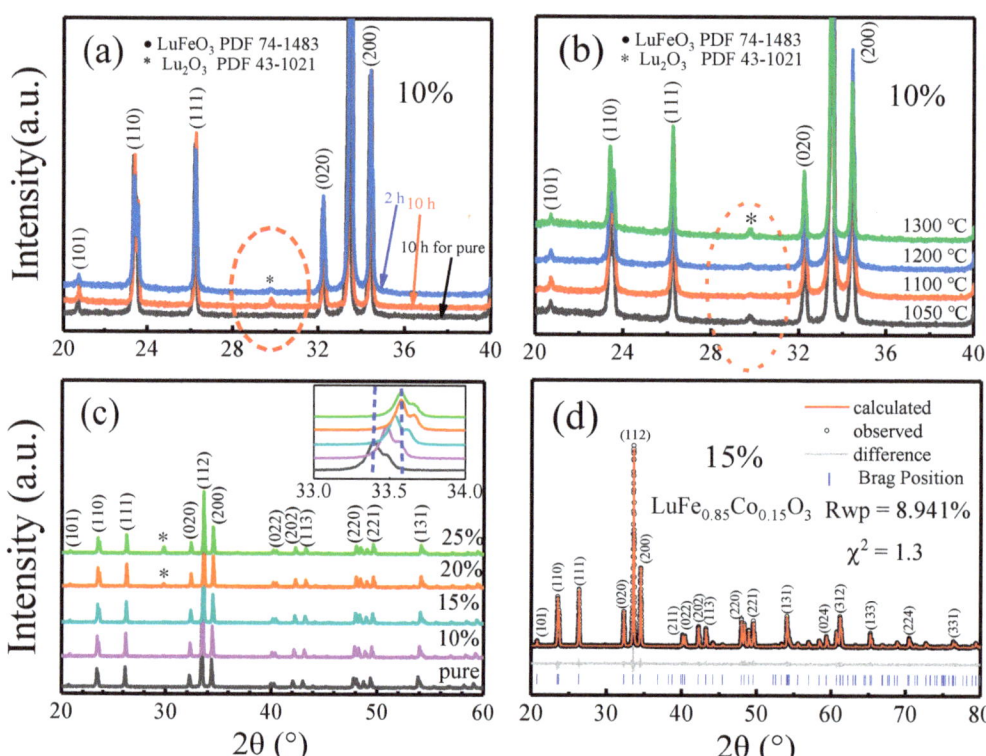

Figure 1. XRD pattern evolution of (**a**) 10% Co doping and calcination time at 1200 °C, (**b**) calcination temperature, and (**c**) Co doping amount (the upper right corner shows the high magnification XRD peak at the (112) plane). (**d**) Refinement of LuFe$_{0.85}$Co$_{0.15}$O$_3$ by TOPAS.

To quantitatively describe the changes in lattice parameters by Co doping, the XRD pattern of a LuFe$_{0.85}$Co$_{0.15}$O$_3$ (LC15) sample with a pure phase was processed by the Rietveld method for refinement using the Topas program, which is a reliable technique that can provide structural details of the materials [34]. The structure was successfully fitted by the perovskite structure with space group Pbnm, as depicted in Figure 1d. All diffraction peaks are well indexed, and no trace of any impurity peak is observed. The weighted profile residual Rwp and goodness of fit χ^2 are 8.941% and 1.3, respectively, indicating a good agreement between the observed and calculated diffraction patterns [35]. All the Rietveld parameters, as obtained from the refinement, plus the counterparts of the parent LuFeO$_3$ (LFO) [33], are tabulated in Table 1. A shrinkage in lattice parameters and volume of LC15 compared with that of LFO is observed. Since the ionic radius of Co^{3+} (54.5 pm in LS and 61.0 pm in HS) is slightly less than that of Fe^{3+} (64.5 pm in HS), shrinkage is expected, indicating a successful substitution of Co ions at Fe sites [17,36,37]. Similar results have also been reported for NdFe$_{1-x}$Co$_x$O$_3$ systems [24–26].

Table 1. Refined structural parameters for LuFe$_{0.85}$Co$_{0.15}$O$_3$ and LuFeO$_3$ samples (error in 10^{-4} order in refined parameters).

Sample Space Group Rwp χ^2	Lattice Parameters (Å)	Atoms	Positions				Volume (Å^3)	Angle Fe-O-Fe
			Wyckoff	x	y	z		
LuFe$_{0.85}$Co$_{0.15}$O$_3$ Pbnm 8.94 1.3	a = 5.1976 ± 1	Lu	4c	0.9800	0.0714	0.25	216.5313 ± 24	
	b = 5.5373 ± 1	Fe(Co)	4b	0	0.5	0		
	c = 7.5235 ± 1	O1	4c	0.1190	0.4539	0.25		140.601
		O2	8d	0.6893	0.3071	0.0621		142.406
LuFeO$_3$ [33] Pbnm 7.74 1.46	a = 5.2154 ± 1	Lu	4c	0.9792	0.0715	0.25	219.0780 ± 35	
	b = 5.5537 ± 1	Fe	4b	0	0.5	0		
	c = 7.5637 ± 1	O1	4c	0.1179	0.4579	0.25		141.632
		O2	8d	0.6890	0.3050	0.0638		142.121

In order to explore the effect of Co doping on the fine structure of the parent LFO, we performed FTIR spectroscopy, an analytical technique capable of revealing the nature of structures by detecting lattice vibrations. The FTIR spectra were recorded in transmission geometry using a KBr disc in the range of 450–3000 cm^{-1}, as illustrated in Figure 2. As per Rao [38], two major bands can be observed at 250–600 cm^{-1}, which are the characteristic bands for rare-earth ferrites. These two bands are attributed to the Fe-O stretching and bending vibrations of the FeO$_6$ octahedral groups in the perovskite compounds [38,39]. In our case, the two bands of LFO are located at 579.2 and 463.3 cm^{-1}, while the counterparts of LC15 blueshift to 585.8 and 478.8 cm^{-1}, respectively. As is well known, the wavenumber of IR bands can be influenced by atomic mass and bond length. However, there is little difference in atomic mass between cobalt (58.9) and iron (55.8) here, so it is reasonable to consider that the bond length of Fe-O is shortened by Co doping, which is consistent with XRD results.

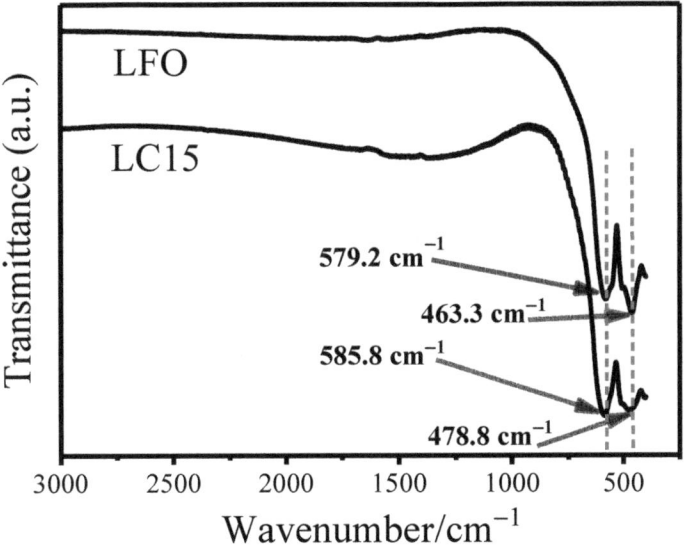

Figure 2. Infrared spectra of the samples LFO and LC15.

2.2. Morphology and Compositional Analysis

Figure 3 presents the surface SEM images, corresponding individual elemental mapping, and EDS spectra results. Figure 2a,e shows the morphologies of LFO and LC15, respectively. It can be seen that the two samples mainly consist of sphere- or polyhedral-shaped particles, and the average particle sizes of the two samples are 1.8 μm and 2.3 μm, respectively. Obviously, Co doping could lead to the growth of the LuFeO$_3$ particles, since a lower reaction temperature and a shorter time of calcination are needed to obtain single-phase Co-doped LuFeO$_3$, according to the XRD results, which are in accordance with the results reported by Somvanshi et al. [27]. The individual elemental distributions of LFO and LC15 are displayed in Figures 3b–d and 3f–i, respectively. It shows that all constituent elements are distributed uniformly. The EDS spectrum of sample LC15 shown in Figure 3j manifests that Lu, Fe, Co, and O elements are obviously contained in the sample, and no other impurities were found. As can be seen from the table in the upper right corner of Figure 3j, the contents of the Fe and Co elements are in line with the nominal ratio.

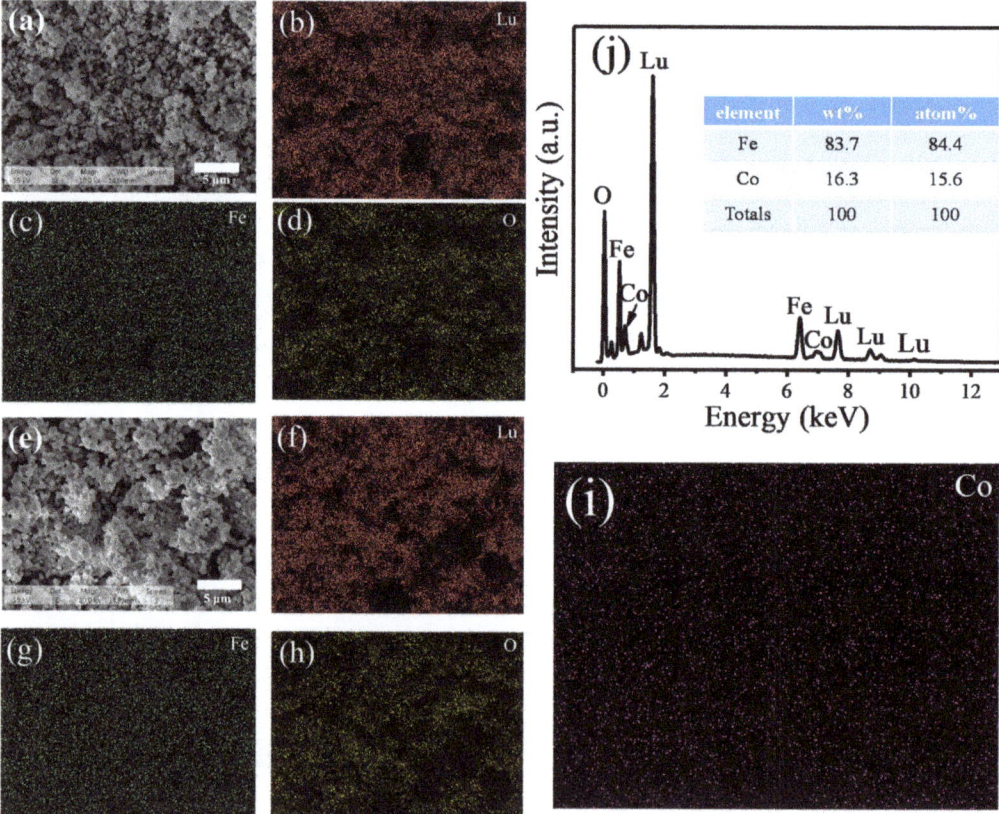

Figure 3. Surface SEM images of LFO (**a**) and LC15 (**e**), and corresponding individual elemental mapping of LFO (**b–d**) and LC15 (**f–i**); (**j**) EDS spectra of LC15 and the elemental relative content of Fe and Co.

2.3. UV–Vis Absorbance Analysis

To evaluate the optical absorbance performance of samples LFO and LC15, the UV–vis absorbance spectra were obtained by scanning from 300 to 800 nm using an integrating sphere attachment. On the one hand, as shown in Figure 4a, the Co-doped LC15 sample exhibits a stronger UV–vis absorption than the pure phase LFO sample, which illustrates that visible light utilization efficiency by Co doping can be improved. On the other hand, the wavelength of the absorption edge extends from ca. 600 nm to ca. 700 nm, which means that Co doping can reduce the bandgap (E_g) and promote the light response range. Based on the above two aspects, the photocatalytic performance should be improved [40].

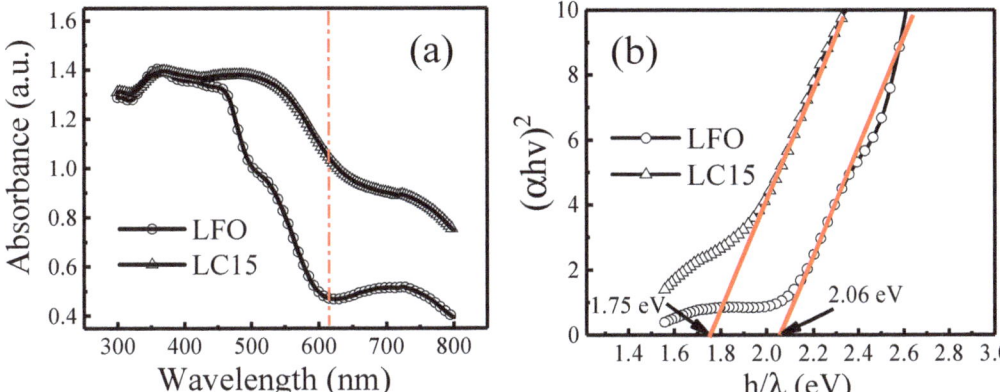

Figure 4. (a) UV–vis absorption spectra of samples LFO and LC15, and (b) corresponding Tauc plot.

To quantitatively analyze the effects of Co doping on E_g, the Tauc relationship below was employed to determine the E_g of these two samples.

$$\alpha h\nu = A(h\nu - E_g)^n \tag{1}$$

where α is the absorption coefficient, $h\nu$ is the photon energy, A is a constant, and E_g is the bandgap [41]. The constant n is determined by the bandgap type of the semiconductor: 0.5 for a direct transition bandgap and 2 for an indirect transition bandgap, where n is 0.5 for LuFeO$_3$ [42]. The approximated E_g was calculated by the straight-line x-intercept in $(\alpha h\nu)^2$ against the $h\nu$ plot transformed from the absorbance spectra, as shown in Figure 4b. The bandgap of the pure phase LFO is 2.06 eV, which is close to the reported value [3,43,44], while the bandgap of LC15 is 1.75 eV. Such a significant decrease in the bandgap was also observed in YbFe$_{1-x}$Co$_x$O$_3$ (x = 0→0.1) and NdFe$_{1-x}$Co$_x$O$_3$ (x = 0→0.4) systems [21,28], in which bandgap decreases are reported to be 2.1→1.72 eV and 2.06→1.46 eV, respectively. These results testified that Co doping could effectively tune the bandgap of orthorhombic rare-earth ferrites. One reason for this is that the B cations in ABO$_3$ play an important role in the bandgap, and replacing B by more electronegative transition elements could reduce the bandgap [45–47]. In this work, the Pauling electronegativity number of Co (1.88) is higher than that of Fe (1.83), which might be responsible for the decrease in the bandgap. Second, the decreased bond length of Fe-O by Co doping enlarges the value of one electron's bandwidth (W). As reported, the increased W decreases the value of the bandgap [48]. Another reason may be due to the weak orbital hybridization of Co 3d and O 2p states, which leads to less binding energy compared to Fe 3d states [21].

2.4. Photocatalytic Degradation of MO under Visible Light Irradiation

To investigate the photocatalytic capacity of Co-doped LuFeO$_3$ particles assisted by OA, experiments were carried out by analyzing the degradation activities of MO solution with an initial concentration of 10 mg/L. OA, a commonly used and environmentally friendly hole scavenger, can be found in most organisms on earth, such as fungi, bacteria, plants, animals, and humans [49]. For an effective photocatalytic process, an optimized catalyst dosage is paramount. Figure 5a shows the change in normalized MO concentration with irradiation time when the dosage of the catalyst LC15 was changed from 0 to 100 mg under the conditions of an initial MO concentration of 10 mg/L and an OA dosage of 12 mg. It can be seen that the MO concentration without a catalyst only decreased by 15% after 5 h of irradiation, while the MO concentration decreased sharply by 64% after adding 10 mg of catalyst. However, as the dosage of catalyst continued to increase, the reduction rates of MO concentration gradually decreased in the cases of 30 and 50 mg of catalyst added. When the dosage of catalyst reached 50 mg, the MO concentration decreased by up to 80%. As the dosage of catalyst continuously increased to 100 mg, the reduction in MO concentration did not increase but rather decreased to 55%. Therefore, 50 mg was considered the optimized catalyst dosage for further photocatalytic experiments. The above phenomenon is reasonable. On the one hand, the intermediate products generated during the photocatalytic reaction process may compete with MO for the active sites on the surface of the catalyst [40], and this competition would become more intense with the increase in the dosage of catalyst. On the other hand, excess catalyst will increase the opacity of the solution, resulting in high light reflectivity, which reduces the absorption of light [50].

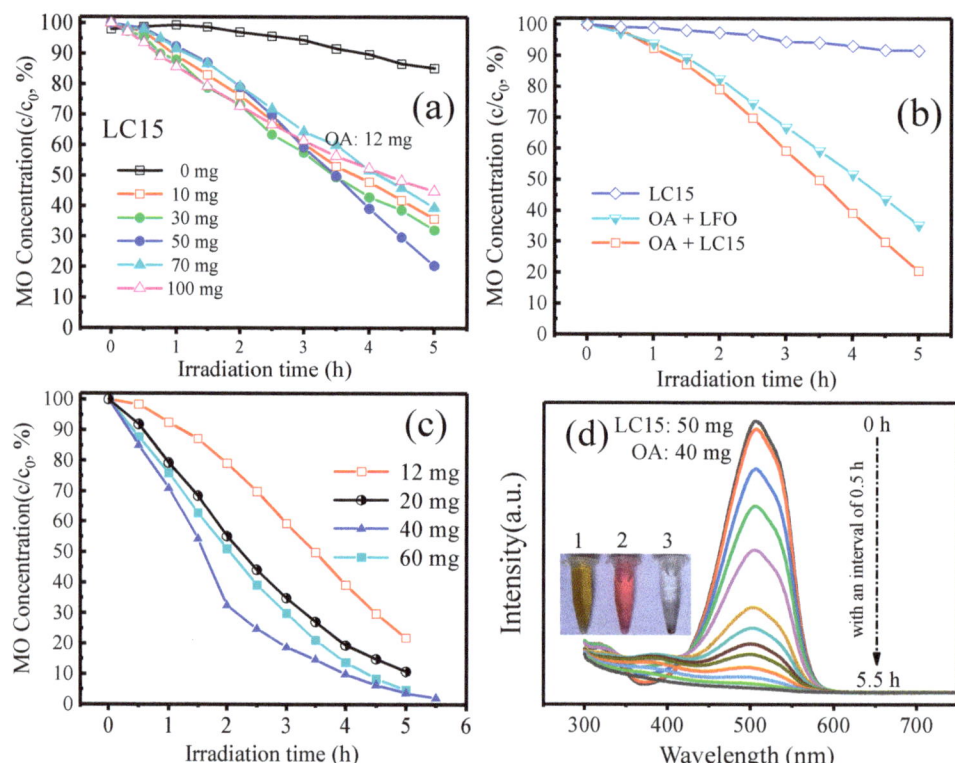

Figure 5. (a) Effects of sample LC15 dosage on the photocatalytic degradation of MO with an OA dosage of 12 mg; (b) comparison of the photocatalytic capacities of sample LC15 (50 mg) and LFO (50 mg); (c) effects of OA dosage on the photocatalytic degradation of MO with an LC15 dosage of 50 mg; (d) evolution of the absorbance spectra of MO of the best results with sample LC15 and OA dosages of 50 mg and 40 mg, respectively. The three centrifugal tubes marked by 1, 2, and 3 denote the MO solution, the MO and OA mixed solution, and the degraded solution after 5.5 h of irradiation.

Figure 5b presents the comparison of the photocatalytic capacities of LFO and LC15 samples. It is observed that the MO concentration only decreased by 8% after 5 h of irradiation for sample LC15 without OA, due to the low photonic efficiency of $LuFeO_3$ resulting from a rapid recombination of photogenerated electron–hole pairs [51]. However, with the assistance of OA, the MO concentration decreased by 65% for the sample LFO. Such a large increase in the photodegradation rate indicates that OA has a great impact on the photodegradation capacity of LuFeO$_3$. In addition, compared with sample LFO, sample LC15 can further reduce the MO concentration by 80%, showing an even stronger photocatalytic capacity. It is well known that the particle size of catalysts has an important effect on photocatalytic performance. Generally speaking, the smaller the particle size, the higher the number of active sites of the catalyst that can participate in photocatalysis and the stronger the catalytic capacity [52]. In our case, the LFO sample has a smaller particle size than the LC15 sample, whereas the photocatalytic performance is weaker than that of the LC15 sample. However, according to the UV–visible absorption results, Co doping can improve the utilization of visible light by increasing the absorption and broadening the spectral response range. Thus, combined with the SEM and UV–visible absorbance observations, it is concluded that the influence of optical properties on the catalytic results is more important than that of particle size in our case [40].

To further optimize the degradation conditions, photocatalytic experiments were conducted by varying the OA dosage from 12 to 60 mg, with the LC15 dosage at 50 mg and the initial MO concentration at 10 mg/L. The results are presented in Figure 5c. It reveals that an excess dosage of OA may lead to the occupation of active sites at the surface of sample LC15, leaving no available sites for photocatalysis [53], and 40 mg of OA is the optimized dosage in our system. The MO concentration can be almost completely degraded (~98%) after 5.5 h of irradiation under the conditions of an LC15 and OA dosage of 50 mg and 40 mg, respectively. Figure 5d displays the evolution of the UV–visible absorption spectra of MO and OA mixed solutions with an interval of 0.5 h after 5.5 h of irradiation, which testifies to the good photocatalytic performance of Co-doped LuFeO$_3$ particles assisted by OA.

2.5. Possible Degradation Mechanism

Based on the above results, a possible degradation mechanism is proposed to explain the enhanced photocatalytic capacity of the Co-doped LuFeO$_3$ sample LC15, as illustrated in Figure 6. Since the redox potentials play a key role in the photocatalytic reaction, it is necessary to determine the effect of Co doping on the conduction band minimum (CBM) and valence band maximum (VBM) potentials of LuFeO$_3$. These energy levels can be calculated by empirical equations [12,54], as follows:

$$E_{CB} = \chi - E^e - 0.5E_g \tag{2}$$

$$E_{VB} = E_{CB} - 0.5E_g \tag{3}$$

where E_{CB} and E_{VB} are the CBM and VBM potentials, E^e is the energy of free electrons versus hydrogen (4.5 eV), and E_g is the bandgap energy. Finally, χ is the electronegativity of a semiconductor, defined as the geometric mean of the electronegativity of the atoms constituting the semiconductor [55], while the electronegativity of an atom is the arithmetic mean [56] of its electron affinity [57,58] and first ionization. The χ value of sample LC15 was estimated to be 5.50, slightly larger than that of sample LFO (5.49) [3]. Thus, the CBM and VBM potentials of LC15 are calculated to be 0.12 and 1.87 eV versus the normal hydrogen electrode (NHE), respectively, as indicated by the orange dotted arrows on the blue scale axis in Figure 6. More recently, Zhou et al. [3] found that ·OH radicals are the dominant active species responsible for the dye's photocatalytic degradation by LuFeO$_3$ particles. Further, in the process of photocatalytic degradation of dye in aqueous solution by a semiconductor, it was found that three redox paths that could produce the free radical ·OH are H_2O/·OH, OH^-/·OH, and O_2/H_2O_2, and their redox potentials are +2.72, +1.89, and +0.695 eV, respectively [59–61]. However, according to the thermodynamic theory, only the O_2/H_2O_2 redox path can occur considering the CBM and VBM potentials of sample LC15, as shown in Figure 6. The related reduction paths are listed below:

$$LC15 + hv \rightarrow LC15(e_{CB}^-) + LC15(h_{VB}^+) \quad (4)$$

$$LC15(2e_{CB}^-) + O_2 + 2H^+ \rightarrow LC15 + H_2O_2 \quad (5)$$

$$LC15(e_{CB}^-) + H_2O_2 \rightarrow LC15 + \cdot OH + OH^- \quad (6)$$

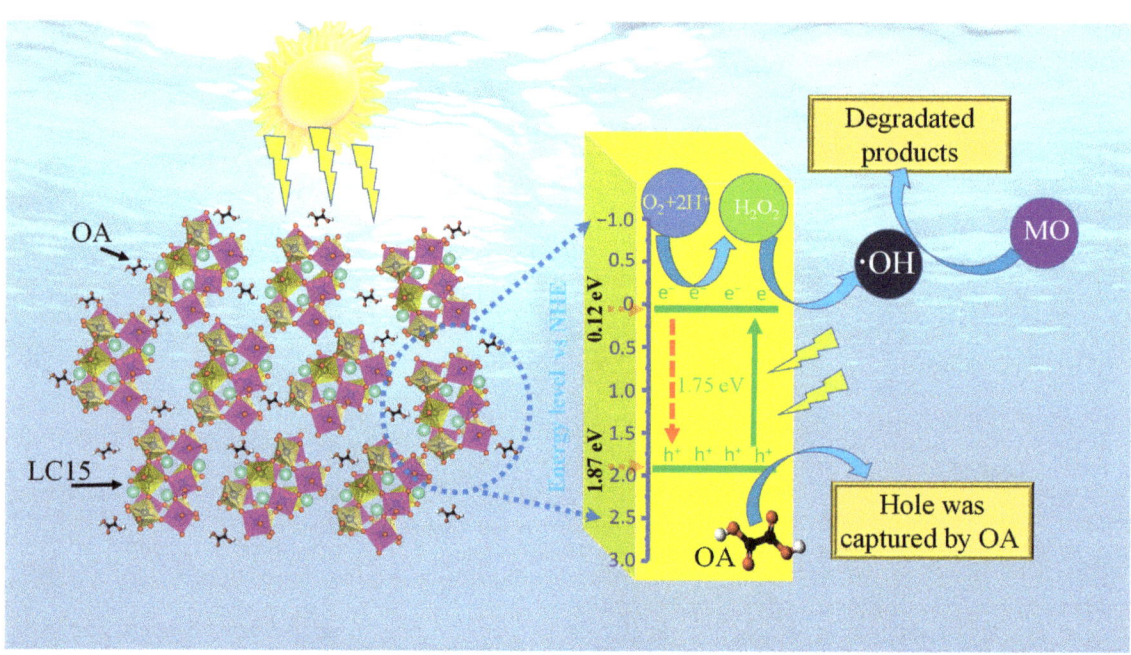

Figure 6. Schematic of the possible mechanism for the photoexcited electron–hole separation and transport processes at the LC15 sample interface, assisted by OA under visible light irradiation.

The oxidation paths of $OH^-/\cdot OH$ are absent in those of sample LC15 compared with the photocatalytic process of LFO [3]. Further, according to the SEM results, the specific surface area of sample LC15 is smaller than that of LFO. However, the degradation efficiency of sample LC15 is still ca. 23% higher than that of LFO. All these results strongly indicate that the enhancement of the light utilization efficiency by Co doping plays a dominant role in improving the photocatalysis.

To reduce the cost of photocatalysis, reusability and stability are two essential features of a photocatalyst. Using the same protocol, the degradation of MO was carried out again by reusing the sample LC15 collected after the first photocatalysis experiment. The comparison of removal efficiency between cycles 1 and 2 is shown in Figure 7a. It is clear that the reduction in removal efficiency is negligible, which suggests that the reusability is good. Figure 7b shows the XRD patterns before and after photocatalysis. Obviously, the orthorhombic structure of sample LC15 is maintained, as all the diffraction peaks corresponding to each plane of orthorhombic $LuFeO_3$ are unchanged and no other diffraction peaks appear. These results indicate the good reusability and stability of sample LC15.

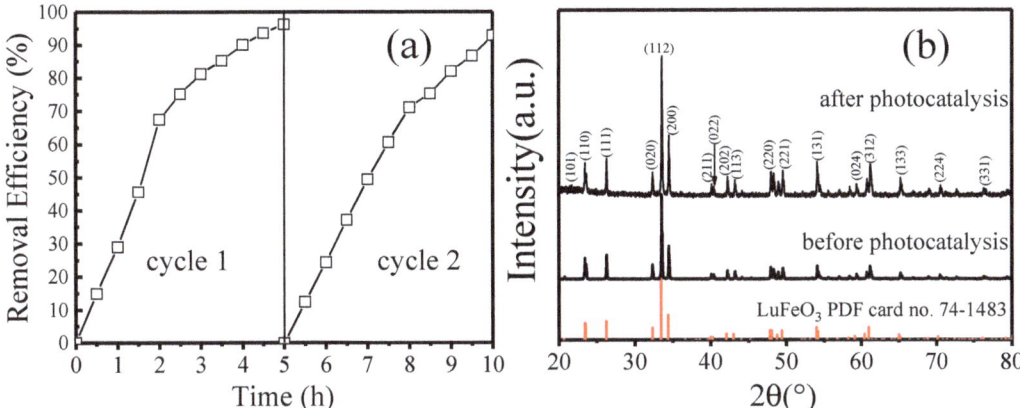

Figure 7. (a) Comparison of the removal efficiency between cycles 1 and 2; (b) XRD patterns of LC15 before and after photocatalysis.

3. Experimental Section

3.1. Chemicals

Highly purified lutetium oxide (Lu_2O_3, 99.99%), ferric sesquioxide (Fe_2O_3, 99.9%), and cobalt oxide (CoO, 99.5%) were purchased from Aladdin Chemistry Co., Ltd. (Shanghai, China). MO ($C_{14}H_{14}N_3NaO_3S$) was purchased from Shanghai Titan Scientific Co., Ltd. (Shanghai, China). OA ($C_2H_2O_4 \cdot 2H_2O$) was purchased from Tianjin Damao Chemical Reagent Factory (Tianjin, China). All the chemicals are of analytical grade or better and were used as received without further purification.

3.2. Preparation of B-Site Co-Doped $LuFeO_3$ Powders

B-site Co-doped $LuFeO_3$ powders were prepared by the MAS method reported in our earlier work [33]. Specifically, 30 mmol of desired stoichiometry of Lu_2O_3, Fe_2O_3, and CoO were slowly added to a 50 mL tungsten carbide–cobalt (WC–Co) pot containing WC-Co milling balls with 30 mL of alcohol. Then, the mixture was blended by a planetary mill machine at a rotational frequency of 586 min^{-1} for 12 h. Next, the mixture with the pot open was transferred to a drying oven and heated at 50 °C for 3 h, and then at 150 °C for 24 h. The dried mixture was sequentially mechanochemically activated at a rotational frequency of 456 min^{-1} for 5 h. Afterward, the powder was calcined at 1000–1300 °C for 2–10 h in a muffle furnace. The final Co-doped $LuFeO_3$ powders were obtained by grounding the calcinated powders thoroughly in an agate crucible.

3.3. Degradation of MO under Visible Light Irradiation

The photocatalytic activities of $LuFeO_3$ (LFO) and $LuFe_{0.85}Co_{0.15}O_3$ (LC15) powders were evaluated by the degradation of MO under visible light irradiation. First, the desired amount of OA was dissolved completely in 100 mL of MO solution (10 mg/L). Second, the mixed solution was transferred to a 250 mL jacketed double-layer glass beaker reactor with the required amount of photocatalysts. After reaching an adsorption–desorption equilibrium by stirring for 30 min in the dark, the photocatalytic reaction under stirring was initiated under the irradiation of a 300 W xenon lamp (MC-PF300, Beijing Merry Change Technology Co., Ltd., Beijing, China) equipped with a 420 nm UV-cut filter positioned 8 cm above the reaction solution. The whole process was operated at 20 °C by circulating thermostatic water and stirring at 450 rpm. During irradiation, a small amount of solution (2 mL) was sampled every 30 min. After the solution was centrifuged at 5000 r/min for 3 min, the supernatant was taken, and its absorbance at 505 nm was measured by a UV–vis spectrophotometer (Agilent Carry 60, Santa Clara, CA, USA) to determine the relative concentration of MO. The normalized MO concentration is defined as $I_t/I_0 \times 100\%$, where I_0 and I_t are the absorbance at 505 nm at the beginning and after a certain time of irradiation.

3.4. Characterization

The phase purity and structure of the as-prepared samples were investigated by a powder XRD instrument (Bruker D8 Advance, Saarbrucken, Germany) using a Ni-filtered Cu Kα (λ = 1.5406 Å) radiation source at 40 kV and 40 mA with a step size of 0.01°. The IR spectroscopy transmittance measurements were performed over a frequency range of 400 to 4000 cm^{-1} using a FTIR spectrometer (Thermal Nicolet Summit, Waltham, MA, USA) with the KBr pellet technique. Morphological and EDS analyses were measured by field-emission SEM (FESEM) (TESCAN Mira4, Brno, Czech). The absorbance measurements were conducted using a UV–visible spectrophotometer (Shimazu SolidSpec-3700, Kyoto, Japan) with an integrating sphere attachment.

4. Conclusions

B-site Co-doped LuFeO$_3$ powders were successfully prepared by a MAS method. The required calcination temperature was lower, and the holding time was shorter than that of the parent phase, LuFeO$_3$. A shrinkage in lattice parameters was observed in Co-doped LuFeO$_3$ samples by fitting the XRD patterns, indicating a successful substitution of Co at Fe sites. FTIR results revealed that two bands attributed to the Fe-O stretching and bending vibrations of the FeO$_6$ octahedral groups both blueshifted in accordance with the XRD results. The particle size of LC15 is slightly larger than that of LFO. The UV–vis absorbance spectra show that Co doping of LuFeO$_3$ can considerably improve the light utilization efficiency by increasing the absorption and broadening the spectral response range. The sample LC15 shows a 23% higher photocatalytic capacity than that of the sample LFO and can decrease the MO concentration by 98% under optimized conditions for 5.5 h of visible light irradiation. A possible degradation mechanism is proposed. In summary, our results clarify that Co doping can improve the photocatalytic performance of orthorhombic LuFeO$_3$.

Author Contributions: Conceptualization, C.S.; Validation, T.S.; Formal analysis, W.X.; Investigation, P.L.; Data curation, W.W.; Writing—review & editing, J.Z.; Supervision, Z.W. All authors have read and agreed to the published version of the manuscript.

Funding: This work was supported by the National Natural Science Foundation of China (Grant No. 61604002), Henan College Key Research Project (No. 20A140013), the Scientific and Technological Project of Anyang City (Grant No. 2021C01SF020), and College Students Innovation Fund of Anyang Normal University (Grant Nos. X202210479121, X202110479097).

Institutional Review Board Statement: Not applicable.

Informed Consent Statement: Not applicable.

Data Availability Statement: Data supporting the results can be provided by the corresponding author via email.

Acknowledgments: The authors sincerely acknowledge Renchun Zhang and Zhi Cao of Anyang Normal University for providing photocatalytic experimental facilitations.

Conflicts of Interest: The authors declare no conflict of interest.

Sample Availability: Not applicable.

References

1. Dong, S.; Xiang, H.; Dagotto, E. Magnetoelectricity in multiferroics: A theoretical perspective. *Natl. Sci. Rev.* **2019**, *6*, 629–641. [CrossRef] [PubMed]
2. Baltz, V.; Manchon, A.; Tsoi, M.; Moriyama, T.; Ono, T.; Tserkovnyak, Y. Antiferromagnetic spintronics. *Rev. Mod. Phys.* **2018**, *90*, 015005. [CrossRef]

3. Zhou, M.; Yang, H.; Xian, T.; Zhang, C.R. A new photocatalyst of LuFeO$_3$ for the dye degradation. *Phys. Scr.* **2015**, *90*, 085808. [CrossRef]
4. Ji, K.; Dai, H.; Deng, J.; Jiang, H.; Zhang, L.; Zhang, H.; Cao, Y. Catalytic removal of toluene over three-dimensionally ordered macroporous Eu$_{1-x}$Sr$_x$FeO$_3$. *Chem. Eng. J.* **2013**, *214*, 262–271. [CrossRef]
5. Humayun, M.; Ullah, H.; Usman, M.; Habibi-Yangjeh, A.; Tahir, A.A.; Wang, C.; Luo, W. Perovskite-type lanthanum ferrite based photocatalysts: Preparation, properties, and applications. *J. Energy Chem.* **2022**, *66*, 314–338. [CrossRef]
6. Niu, X.; Du, W.; Du, W. Preparation, characterization and gas-sensing properties of rare earth mixed oxides. *Sens. Actuators B-Chem.* **2004**, *99*, 399–404. [CrossRef]
7. Bartel, C.J.; Sutton, C.; Goldsmith, B.R.; Ouyang, R.; Musgrave, C.B.; Ghiringhelli, L.M.; Scheffler, M. New tolerance factor to predict the stability of perovskite oxides and halides. *Sci. Adv.* **2019**, *5*, eaav0693. [CrossRef]
8. Tokunaga, Y.; Taguchi, Y.; Arima, T.-H.; Tokura, Y. Electric-field-induced generation and reversal of ferromagnetic moment in ferrites. *Nat. Phys.* **2012**, *8*, 838–844. [CrossRef]
9. Phan, T.T.N.; Nikoloski, A.N.; Bahri, P.A.; Li, D. Heterogeneous photo-Fenton degradation of organics using highly efficient Cu-doped LaFeO$_3$ under visible light. *J. Ind. Eng. Chem.* **2018**, *61*, 53–64. [CrossRef]
10. Ju, L.; Tang, X.; Zhang, Y.; Li, X.; Cui, X.; Yang, G. Single Selenium Atomic Vacancy Enabled Efficient Visible-Light-Response Photocatalytic NO Reduction to NH$_3$ on Janus WSSe Monolayer. *Molecules* **2023**, *28*, 2959. [CrossRef]
11. Ju, L.; Tang, X.; Li, J.; Dong, H.; Yang, S.; Gao, Y.; Liu, W. Armchair Janus WSSe Nanotube Designed with Selenium Vacancy as a Promising Photocatalyst for CO$_2$ Reduction. *Molecules* **2023**, *28*, 4602. [CrossRef] [PubMed]
12. Nethercot, A.H. Prediction of Fermi Energies and Photoelectric Thresholds Based on Electronegativity Concepts. *Phys. Rev. Lett.* **1974**, *33*, 1088–1091. [CrossRef]
13. White, R.L. Review of Recent Work on the Magnetic and Spectroscopic Properties of the Rare-Earth Orthoferrites. *J. Appl. Phys.* **1969**, *40*, 1061–1069. [CrossRef]
14. Yamaguchi, T. Theory of spin reorientation in rare-earth orthochromites and orthoferrites. *J. Phys. Chem. Solids* **1974**, *35*, 479–500. [CrossRef]
15. Zhu, W.; Pi, L.; Tan, S.; Zhang, Y. Anisotropy and extremely high coercivity in weak ferromagnetic LuFeO$_3$. *Appl. Phys. Lett.* **2012**, *100*, 052407. [CrossRef]
16. Sun, H.H.; Kim, U.-H.; Park, J.-H.; Park, S.-W.; Seo, D.-H.; Heller, A.; Mullins, C.B.; Yoon, C.S.; Sun, Y.-K. Transition metal-doped Ni-rich layered cathode materials for durable Li-ion batteries. *Nat. Commun.* **2021**, *12*, 6552. [CrossRef] [PubMed]
17. Poudel, M.B.; Ojha, G.P.; Kim, A.A.; Kim, H.J. Manganese-doped tungsten disulfide microcones as binder-free electrode for high performance asymmetric supercapacitor. *J. Energy Storage* **2022**, *47*, 103674. [CrossRef]
18. Arshad, Y.; Khan, S.; Hashmi, M.A.; Ayub, M. Transition metal doping: A new and effective approach for remarkably high nonlinear optical response in aluminum nitride nanocages. *New J. Chem.* **2018**, *42*, 6976–6989. [CrossRef]
19. Zhang, W.; Xi, R.; Li, Y.; Zhang, Y.; Wang, P.; Hu, D. Recent development of transition metal doped carbon materials derived from biomass for hydrogen evolution reaction. *Int. J. Hydrogen Energy* **2022**, *47*, 32436–32454. [CrossRef]
20. Poudel, M.B.; Kim, H.J. Synthesis of high-performance nickel hydroxide nanosheets/gadolinium doped-α-MnO$_2$ composite nanorods as cathode and Fe$_3$O$_4$/GO nanospheres as anode for an all-solid-state asymmetric supercapacitor. *J. Energy Chem.* **2022**, *64*, 475–484. [CrossRef]
21. Polat, O.; Caglar, M.; Coskun, F.M.; Coskun, M.; Caglar, Y.; Turut, A. An experimental investigation: The impact of cobalt doping on optical properties of YbFeO$_{3-δ}$ thin film. *Mater. Res. Bull.* **2019**, *119*, 110567. [CrossRef]
22. Ishihara, T.; Furutani, H.; Honda, M.; Yamada, T.; Shibayama, T.; Akbay, T.; Sakai, N.; Yokokawa, H.; Takita, Y. Improved Oxide Ion Conductivity in La$_{0.8}$Sr$_{0.2}$Ga$_{0.8}$Mg$_{0.2}$O$_3$ by Doping Co. *Chem. Mater.* **1999**, *11*, 2081–2088. [CrossRef]
23. Feng, G.; Xue, Y.; Shen, H.; Feng, S.; Li, L.; Zhou, J.; Yang, H.; Xu, D. Sol–gel synthesis, solid sintering, and thermal stability of single-phase YCoO$_3$. *Phys. Status Solidi A* **2012**, *209*, 1219–1224. [CrossRef]
24. Zhang, R.; Hu, J.; Han, Z.; Zhao, M.; Wu, Z.; Zhang, Y.; Qin, H. Electrical and CO-sensing properties of NdFe$_{1-x}$Co$_x$O$_3$ perovskite system. *J. Rare Earth* **2010**, *28*, 591–595. [CrossRef]
25. Somvanshi, A.; Husain, S.; Khan, W. Investigation of structure and physical properties of cobalt doped nano-crystalline neodymium orthoferrite. *J. Alloys Compd.* **2019**, *778*, 439–451. [CrossRef]
26. Nforna, E.A.; Tsobnang, P.K.; Fomekong, R.L.; Tedjieukeng, H.M.K.; Lambi, J.N.; Ghogomu, J.N. Effect of B-site Co substitution on the structure and magnetic properties of nanocrystalline neodymium orthoferrite synthesized by auto-combustion. *R. Soc. Open Sci.* **2021**, *8*, 201883. [CrossRef]
27. Somvanshi, A.; Husain, S.; Manzoor, S.; Zarrin, N.; Ahmad, N.; Want, B.; Khan, W. Tuning of magnetic properties and multiferroic nature: Case study of cobalt-doped NdFeO$_3$. *Appl. Phys. A* **2021**, *127*, 174. [CrossRef]
28. Nguyen, T.A.; Pham, T.L.; Mittova, I.Y.; Mittova, V.O.; Nguyen, T.L.T.; Nguyen, H.V.; Bui, V.X. Co-Doped NdFeO$_3$ Nanoparticles: Synthesis, Optical, and Magnetic Properties Study. *Nanomaterials* **2021**, *11*, 937. [CrossRef] [PubMed]
29. Polat, O.; Coskun, M.; Coskun, F.M.; Zlamal, J.; Kurt, B.Z.; Durmus, Z.; Caglar, M.; Turut, A. Co doped YbFeO$_3$: Exploring the electrical properties via tuning the doping level. *Ionics* **2019**, *25*, 4013–4029. [CrossRef]
30. Qahtan, A.A.A.; Husain, S.; Somvanshi, A.; Fatema, M.; Khan, W. Investigation of alteration in physical properties of dysprosium orthochromite instigated through cobalt doping. *J. Alloys Compd.* **2020**, *843*, 155637. [CrossRef]

31. Haque, A.; Bhattacharya, S.; Das, R.; Hossain, A.; Gayen, A.; Kundu, A.K.; Vasundhara, M.; Seikh, M.M. Effects of Bi doping on structural and magnetic properties of cobalt ferrite perovskite oxide LaCo$_{0.5}$Fe$_{0.5}$O$_3$. *Ceram. Int.* **2022**, *48*, 16348–16356. [CrossRef]
32. Wang, Z.-Q.; Lan, Y.-S.; Zeng, Z.-Y.; Chen, X.-R.; Chen, Q.-F. Magnetic structures and optical properties of rare-earth orthoferrites RFeO$_3$ (R=Ho, Er, Tm and Lu). *Solid State Commun.* **2019**, *288*, 10–17. [CrossRef]
33. Wang, Z.; Xiao, W.; Zhang, J.; Huang, J.; Dong, M.; Yuan, H.; Xu, T.; Shi, L.; Dai, Y.; Liu, Q.; et al. Effects of mechanochemical activation on the structural and electrical properties of orthorhombic LuFeO$_3$ ceramics. *J. Am. Ceram. Soc.* **2021**, *104*, 3019–3029. [CrossRef]
34. Hill, R.J.; Howard, C.J. Quantitative phase analysis from neutron powder diffraction data using the Rietveld method. *J. Appl. Crystallogr.* **1987**, *20*, 467–474. [CrossRef]
35. Young, R.A. *The Rietveld Method*; Oxford University Press: London, UK, 1995; p. 308.
36. Shannon, R. Revised effective ionic radii and systematic studies of interatomic distances in halides and chalcogenides. *Acta Crystallogr. A* **1976**, *32*, 751–767. [CrossRef]
37. Baloch, A.A.B.; Alqahtani, S.M.; Mumtaz, F.; Muqaibel, A.H.; Rashkeev, S.N.; Alharbi, F.H. Extending Shannon's ionic radii database using machine learning. *Phys. Rev. Mater.* **2021**, *5*, 043804. [CrossRef]
38. Rao, G.V.S.; Rao, C.N.R.; Ferraro, J.R. Infrared and Electronic Spectra of Rare Earth Perovskites: Ortho-Chromites, -Manganites and -Ferrites. *Appl. Spectrosc.* **1970**, *24*, 436–445. [CrossRef]
39. Kaczmarek, W.; Pajak, Z.; Połomska, M. Differential thermal analysis of phase transitions in (Bi$_{1-x}$La$_x$)FeO$_3$ solid solution. *Solid State Commun.* **1975**, *17*, 807–810. [CrossRef]
40. Govindan, K.; Chandran, H.T.; Raja, M.; Maheswari, S.U.; Rangarajan, M. Electron scavenger-assisted photocatalytic degradation of amido black 10B dye with Mn$_3$O$_4$ nanotubes: A response surface methodology study with central composite design. *J. Photochem. Photobiol. A Chem.* **2017**, *341*, 146–156. [CrossRef]
41. Tauc, J.; Grigorovici, R.; Vancu, A. Optical Properties and Electronic Structure of Amorphous Germanium. *Phys. Status Solidi B* **1966**, *15*, 627–637. [CrossRef]
42. Zhu, L.P.; Deng, H.M.; Sun, L.; Yang, J.; Yang, P.X.; Chu, J.H. Optical properties of multiferroic LuFeO$_3$ ceramics. *Ceram. Int.* **2014**, *40 Pt A*, 1171–1175. [CrossRef]
43. Adams, D.J.; Amadon, B. Study of the volume and spin collapse in orthoferrite LuFeO$_3$ using LDA+U. *Phys. Rev. B* **2009**, *79*, 115114. [CrossRef]
44. Zhou, M.; Yang, H.; Xian, T.; Ma, J.Y.; Zhang, H.M.; Feng, W.J.; Wei, Z.Q.; Jiang, J.L. Morphology-controlled synthesis of orthorhombic LuFeO$_3$ particles via a hydrothermal route. *J. Alloys Compd.* **2014**, *617*, 855–862. [CrossRef]
45. Polat, O.; Durmus, Z.; Coskun, F.M.; Coskun, M.; Turut, A. Engineering the band gap of LaCrO$_3$ doping with transition metals (Co, Pd, and Ir). *J. Mater. Sci.* **2018**, *53*, 3544–3556. [CrossRef]
46. Polat, O.; Coskun, M.; Coskun, F.M.; Durmus, Z.; Caglar, M.; Turut, A. Os doped YMnO$_3$ multiferroic: A study investigating the electrical properties through tuning the doping level. *J. Alloys Compd.* **2018**, *752*, 274–288. [CrossRef]
47. Polat, O.; Coskun, M.; Coskun, F.M.; Zengin Kurt, B.; Durmus, Z.; Caglar, Y.; Caglar, M.; Turut, A. Electrical characterization of Ir doped rare-earth orthoferrite YbFeO$_3$. *J. Alloys Compd.* **2019**, *787*, 1212–1224. [CrossRef]
48. Rhaman, M.M.; Matin, M.A.; Hossain, M.N.; Mozahid, F.A.; Hakim, M.A.; Islam, M.F. Bandgap engineering of cobalt-doped bismuth ferrite nanoparticles for photovoltaic applications. *Bull. Mater. Sci.* **2019**, *42*, 190. [CrossRef]
49. Palmieri, F.; Estoppey, A.; House, G.L.; Lohberger, A.; Bindschedler, S.; Chain, P.S.G.; Junier, P. Chapter Two—Oxalic acid, a molecule at the crossroads of bacterial-fungal interactions. *Adv. Appl. Microbiol.* **2019**, *106*, 49–77.
50. Kuriechen, S.K.; Murugesan, S.; Raj, S.P.; Maruthamuthu, P. Visible light assisted photocatalytic mineralization of Reactive Red 180 using colloidal TiO$_2$ and oxone. *Chem. Eng. J.* **2011**, *174*, 530–538. [CrossRef]
51. Polat, O.; Coskun, F.M.; Yildirim, Y.; Sobola, D.; Ercelik, M.; Arikan, M.; Coskun, M.; Sen, C.; Durmus, Z.; Caglar, Y.; et al. The structural studies and optical characteristics of phase-segregated Ir-doped LuFeO$_{3-\delta}$ films. *Appl. Phys. A* **2023**, *129*, 198. [CrossRef]
52. Tomkiewicz, M. Scaling properties in photocatalysis. *Catal. Today* **2000**, *58*, 115–123. [CrossRef]
53. Zhang, G.; Zhang, W.; Crittenden, J.; Minakata, D.; Chen, Y.; Wang, P. Effects of inorganic electron donors in photocatalytic hydrogen production over Ru/(CuAg)$_{0.15}$In$_{0.3}$Zn$_{1.4}$S$_2$ under visible light irradiation. *J. Renew. Sustain. Energy* **2014**, *6*, 033131. [CrossRef]
54. Morrison, S.R. *Electrochemistry at Semiconductor and Oxidized Metal Electrodes*; Springer: New York, NY, USA, 1980; p. 183.
55. Yuan, Q.; Chen, L.; Xiong, M.; He, J.; Luo, S.-L.; Au, C.-T.; Yin, S.-F. Cu$_2$O/BiVO$_4$ heterostructures: Synthesis and application in simultaneous photocatalytic oxidation of organic dyes and reduction of Cr(VI) under visible light. *Chem. Eng. J.* **2014**, *255*, 394–402. [CrossRef]
56. Pearson, R.G. Absolute electronegativity and absolute hardness of Lewis acids and bases. *J. Am. Chem. Soc.* **1985**, *107*, 6801–6806. [CrossRef]
57. Andersen, T.; Haugen, H.K.; Hotop, H. Binding Energies in Atomic Negative Ions: III. *J. Phys. Chem. Ref. Data* **1999**, *28*, 1511–1533. [CrossRef]
58. Ning, C.; Lu, Y. Electron Affinities of Atoms and Structures of Atomic Negative Ions. *J. Phys. Chem. Ref. Data* **2022**, *51*, 021502. [CrossRef]

59. Jiang, H.-Y.; Liu, J.; Cheng, K.; Sun, W.; Lin, J. Enhanced Visible Light Photocatalysis of Bi_2O_3 upon Fluorination. *J. Phys. Chem. C* **2013**, *117*, 20029–20036. [CrossRef]
60. Tachikawa, T.; Fujitsuka, M.; Majima, T. Mechanistic Insight into the TiO_2 Photocatalytic Reactions: Design of New Photocatalysts. *J. Phys. Chem. C* **2007**, *111*, 5259–5275. [CrossRef]
61. Jiang, H.-Y.; Cheng, K.; Lin, J. Crystalline metallic Au nanoparticle-loaded α-Bi_2O_3 microrods for improved photocatalysis. *Phys. Chem. Chem. Phys.* **2012**, *14*, 12114–12121. [CrossRef]

Disclaimer/Publisher's Note: The statements, opinions and data contained in all publications are solely those of the individual author(s) and contributor(s) and not of MDPI and/or the editor(s). MDPI and/or the editor(s) disclaim responsibility for any injury to people or property resulting from any ideas, methods, instructions or products referred to in the content.

Article

Armchair Janus WSSe Nanotube Designed with Selenium Vacancy as a Promising Photocatalyst for CO_2 Reduction

Lin Ju [1,*], Xiao Tang [2], Jingli Li [1], Hao Dong [3], Shenbo Yang [4], Yajie Gao [1] and Wenhao Liu [1]

[1] School of Physics and Electric Engineering, Anyang Normal University, Anyang 455000, China; 201104015@stu.aynu.edu.cn (J.L.); 201101047@stu.aynu.edu.cn (Y.G.)
[2] Institute of Materials Physics and Chemistry, College of Science, Nanjing Forestry University, Nanjing 210037, China; xiaotang@njfu.edu.cn
[3] College of Physical Science and Technology, Central China Normal University, Wuhan 430079, China; dh123456@mails.ccnu.edu.cn
[4] Hongzhiwei Technology (Shanghai) Co., Ltd., 1599 Xinjinqiao Road, Pudong, Shanghai 200120, China; yangshenbo@hzwtech.com
* Correspondence: julin@aynu.edu.cn

Citation: Ju, L.; Tang, X.; Li, J.; Dong, H.; Yang, S.; Gao, Y.; Liu, W. Armchair Janus WSSe Nanotube Designed with Selenium Vacancy as a Promising Photocatalyst for CO_2 Reduction. *Molecules* 2023, 28, 4602. https://doi.org/10.3390/molecules28124602

Academic Editor: Sergio Navalon

Received: 15 May 2023
Revised: 30 May 2023
Accepted: 5 June 2023
Published: 7 June 2023

Copyright: © 2023 by the authors. Licensee MDPI, Basel, Switzerland. This article is an open access article distributed under the terms and conditions of the Creative Commons Attribution (CC BY) license (https://creativecommons.org/licenses/by/4.0/).

Abstract: Photocatalytic conversion of carbon dioxide into chemical fuels offers a promising way to not only settle growing environmental problems but also provide a renewable energy source. In this study, through first-principles calculation, we found that the Se vacancy introduction can lead to the transition of physical-to-chemical CO_2 adsorption on Janus WSSe nanotube. Se vacancies work at the adsorption site, which significantly improves the amount of transferred electrons at the interface, resulting in the enhanced electron orbital hybridization between adsorbents and substrates, and promising the high activity and selectivity for carbon dioxide reduction reaction (CO_2RR). Under the condition of illumination, due to the adequate driving forces of photoexcited holes and electrons, oxygen generation reaction (OER) and CO_2RR can occur spontaneously on the S and Se sides of the defective WSSe nanotube, respectively. The CO_2 could be reduced into CH_4, meanwhile, the O_2 is produced by the water oxidation, which also provides the hydrogen and electron source for the CO_2RR. Our finding reveals a candidate photocatalyst for obtaining efficient photocatalytic CO_2 conversion.

Keywords: CO_2 reduction; Se vacancy; photocatalysis; Janus WSSe nanotube

1. Introduction

For the last few years, given the limitation of fossil fuel reserves and the growth of atmospheric CO_2 levels, an urgent need has existed to create a sustainable option for converting unwanted CO_2 into useful products in the form of chemicals and fuels [1–3], which will not only solve the greenhouse effect, melting glaciers, and other environmental problems caused by carbon dioxide, but also alleviate the current energy crisis [4]. The conversion of carbon dioxide could be operated through a variety of pathways, including biochemical [5], electrochemical [6,7], photochemical [8,9], and thermochemical [10] reactions. As sunlight is a theoretically unlimited power source, solar-powered CO_2 reduction can be perceived as the best option among these promising approaches [11,12]. Until now, photocatalytic CO_2RR has attracted great attentions and achieved many results [13–16]. Photocatalysis is widely believed to have three primary key steps, i.e., sunlight harvesting by the semiconductor (hν > Eg), photo-generated carrier separation and transport, and reactions on the surface [17–20]. While many solar active catalysts for CO_2 photoreduction have been reported, they mostly suffer from instability, poor energy conversion rates, non-controllable selectivity, and failure to fully inhibit competing hydrogen evolution reactions (HER) in existence with water [21,22]. Consequently, it remains a great priority to design high-activity photocatalysts for CO_2 reduction with great conversion efficiency and selectivity.

Soon after the Janus-structured MoSSe monolayer was fabricated by a modified chemical vapor deposition (CVD) method based on the sulfidation of $MoSe_2$ monolayer [23] and the selenization of the MoS_2 monolayer [24], the two-dimensional (2D) Janus transition metal dichalcogenides, such as Janus MoSSe and WSSe, have become candidates with great potential application for photocatalysis because of their excellent optical absorption, suitable band edge positions, and high carrier separation [19,23–26]. Our previous work demonstrated tha, the tubular Janus WSSe, obtained by rolling the planar Janus WSSe with an acceptable strain energy, possesses an enhanced electrostatic potential difference between the Se and S layers, resulting in a stronger built-in electric field than the planar structure. The stronger built-in electric field usually could help to strengthen the adsorption of small gas molecules, even to activate them. For CO_2RR, due to the inertness of the CO_2 gas molecule induced by the strong C=O bonds, effective activation of the CO_2 molecule is key for the subsequent reduction. Therefore, to explore the photocatalytic CO_2RR performance of the Janus WSSe nanotube is meaningful for developing a highly efficient photocatalyst.

In our research, the CO_2 adsorption on the Janus WSSe nanotube in pristine and defective states had been studied using DFT calculations. Adsorption energy (E_{ads}), charge density difference (CDD), and density of state (DOS) were employed to explain the coupling between the substrate and adsorbate. It was found that the introduction of Se vacancies on Janus WSSe could brilliantly change the physical adsorption of CO_2 into chemical adsorption, which effectively activates the CO_2 gas molecules and makes CO_2RR possible. The semiconducting property of the defective Janus WSSe nanotube is confirmed by the electronic band structure. We studied its photocatalytic CO_2RR performance by analyzing the absorption spectrum, redox capacity, and reaction driving force of photo-excited carriers. Furthermore, in order to keep the CO_2RR sustained and stable, we also consider the OER reaction on the S side of the defective Janus WSSe nanotube. Furthermore, competition from the CO_2RR and hydrogen evolution reaction (HER) is addressed. We found that the defective WSSe nanotubes have excellent photocatalytic properties and can serve as a hopeful photocatalyst for light-driven CO_2 reduction.

2. Results and Discussion

2.1. The CO_2 Adsorption on Pristine Janus WSSe Nanotube

Janus WSSe nanotubes are constructed by scrolling Janus WSSe monolayers, whereby the W layer is interposed between the Se and S layers. Our previous work reported that the strain energy for the formation (0.10 eV/atom) of the Janus WSSe nanotubes with a structure of Se layer on the outside and S layer on the inside is lower than the one (0.23 eV/atom) with a contrary structure, indicating relatively more stability [27]. Herein, we chose the (12, 12) armchair Janus WSSe nanotube as the substrate in the adsorption system. As shown in Figure S1, the diameter is 21.86 Å and the height of Se-S is 3.22 Å. The W-S and W-Se bond lengths are 2.38 and 2.60 Å, respectively, separately a little shorter and larger than the corresponding ones (2.41 and 2.52 Å) in the planar structure [28]. In this study, we only considered the CO_2 adsorption on the outer side (Se side) of the nanotube, and the case of the adsorption on the inner side is neglected because the CO_2 gas molecules are difficult to pass through the nanotube walls to arrive on the inner side (the barrier is up to 28.33 eV, see Figure S2). We put a CO_2 gas molecule on the Se side of the nanotube to build the adsorption system and completely relax it. As shown in Figure 1, there are four adsorption sites taken into consideration, namely **center** (above center of the hexagon), **bond** (above W-Se bond), and **W/Se** (above W/Se atom).

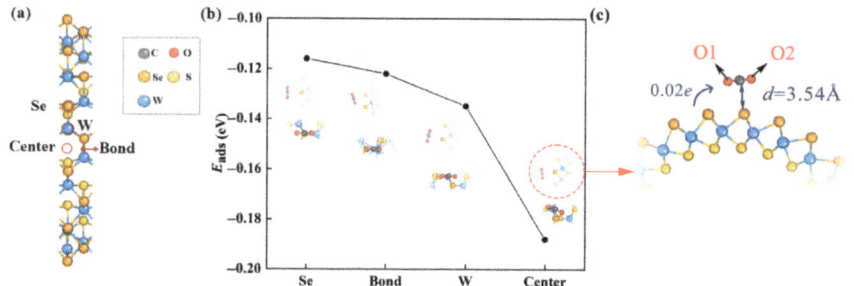

Figure 1. (a) The adsorption sites (marked with red circles) in consideration of the pristine Janus WSSe nanotube. (b) The adsorption energy as well as the top (upper) and side (lower) views of the optimized configurations of CO_2 gas molecule adsorbing on pristine WSSe nanotube with different adsorption sites. The gray, red, orange, yellow, and blue balls each represent C, O, Se, S, and W atoms. (c) The enlarged view for the top view of center adsorption site optimized structure. The adsorption distance between the substrate and the adsorbate is represented by the dark blue, d. Transfer of charge from substrate to CO_2 molecule identified by the blue arrow.

According to Equation (1), we obtained E_{ads} values of various adsorption sites, which were used to explore the most stable adsorption configuration. As shown in Figure 1b, we found that the E_{ads} arrived the smallest (−0.19 eV) when the adsorbed CO_2 gas molecule was located at the **center** site, which was the most stable adsorption configuration. The small absolute value of E_{ads} of this adsorption configuration revealed that the adsorption is physical adsorption (usually, $|E_{ads}| \leq 1$ eV [29–32]).

We studied the mechanism of the CO_2 physisorption on the pristine Janus WSSe nanotube in detail based on the adsorption distance and Bader charge results. The CO_2 gas molecule kept the linear morphology after adsorption (see Figure 1c), and the distance from the C atom of the CO_2 gas molecule to its nearest Se atom of the pristine Janus WSSe nanotube is as high as 3.54 Å, which greatly exceeds the Se-C bond length (2.29 Å). In addition, the amount of transfer electron, moving from the pristine Janus WSSe nanotube to the CO_2 molecule, is only 0.02 e, indicating the weak interaction between the substrate and the CO_2 molecule.

At the same time, we also calculated the DOS values of the adsorption configurations. As can be seen from Figure 2a–c, the projected DOS of the WSSe nanotube has a negligible change compared with those of the corresponding pristine WSSe nanotube, indicating that the electronic properties of the WSSe nanotube remain. However, there is a significant difference of the DOS between the adsorbed gas molecule and the pristine gas molecule, which is due to charge rearrangement after adsorption; that is, the O atoms gain electrons, while the C atom loses electrons (as listed in Table S1). The little orbital hybridization between the WSSe nanotube and CO_2, mainly composed of the Se p and CO_2 O p orbitals, is consistent with the tiny interfacial electron transfer, demonstrating that the interaction between the WSSe nanotube and molecules is weak. According to the above analysis, it can be determined that the adsorption of CO_2 by the pristine WSSe nanotube is physisorption.

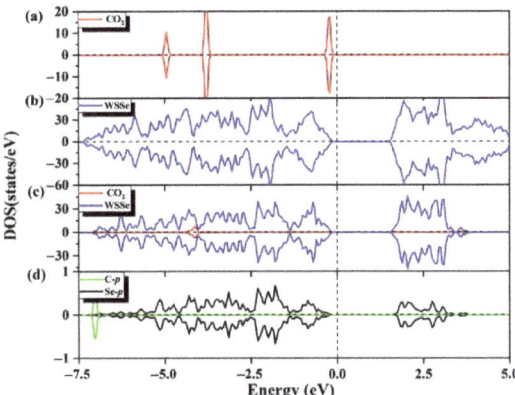

Figure 2. (a) The total state density of the pristine CO_2 gas molecules and (b) the total state density of the pristine WSSe nanotube. (c) The partial state density of the adsorption system, where WSSe nanotube is shown in dark blue and CO_2 is shown in red. (d) Partial state densities of the C p orbitals (cyan) of the adsorbed CO_2 gas molecule and the Se p orbitals (black) of the Se atom most nearby the adsorbed CO_2 molecule. Fermi level is expressed by the vertical dashed line.

2.2. The CO_2 Adsorption on Defective Janus Wsse Nanotube

The pristine WSSe nanotube can be used as a gas collection system for physical CO_2 adsorption. However, in order to convert the CO_2 gas into value-added industrial raw materials through chemical reactions, chemical adsorption of CO_2 is required, which requires the substrate to have a stronger adsorption capacity. Our earlier results have reported that introducing vacancy defects could effectively improve the stability of the geometric structures for some gas adsorption systems, making the adsorption capacity of the substrate increase [33–35].

Since the CO_2 is more easily adsorbed on the Se side of WSSe nanotube, hereby, we applied the Se vacancy defects into the Janus WSSe nanotube to enhance its CO_2 adsorption capacity, which also has been demonstrated to be more easily formed than the S and W vacancy defects in the WSSe layered material [33]. Based on the analysis on the elastic modulus, we find that a low Se vacancy concentration does not affect the mechanical property of the Janus WSSe nanotube drastically. (More details can be found in the Supporting Information and Figure S3). The calculated E_{ads} of CO_2 molecule adsorbing on defective Janus WSSe nanotube is −1.41 eV, greatly exceeding the one (−0.19 eV) on the pristine Janus WSSe nanotube, indicating that the introduction of Se vacancy strengthens the CO_2 adsorption. More interesting, as displayed in Figure 3a, the adsorbed CO_2 molecule undergoes an obvious deformation from the initial linear shape into the bending one (∠OCO = 114.17°). Additionally, one of the C=O bonds in the adsorbed CO_2 molecule (C-O2 bond) transforms into the C-O bond, and the C and O2 atoms bond to different W atoms, respectively. The obvious deformation demonstrates that the CO_2 molecule could be activated by the defective Janus WSSe nanotube. However, the defective planar WSSe monolayer does not have such high activity. The adsorbed CO_2 molecule on the defective planar WSSe monolayer keeps its linear shape (see Figure S4), and the adsorption energy in this case is only −0.20 eV. This phenomenon can be explained by the following reasons: (I) bending the planar structure allows more of the W atom area to be exposed, enlarging the contact surface of the CO_2 molecule on the W atom; (II) the W atoms near the Se vacancy in the tubular structure WSSe have more electrons (0.15 e/atom) than the ones in the planar structure, according to the Bader charge results, which leads to easier electron transfer from W atoms on the WSSe nanotubes to the CO_2 molecule and facilitates the formation of strong bonds.

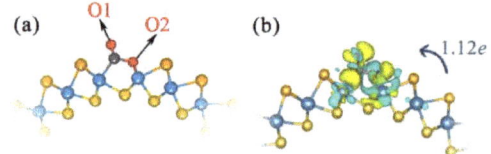

Figure 3. Top view (**a**) of the optimized structure and CDD (**b**) of Janus WSSe nanotube with Se vacancy adsorbed CO_2 gas molecules. Cyan (yellow) areas indicate charge depletion (accumulation). The isosurface level is 0.002 $e\text{Å}^{-3}$. Transfer of charge from substrate to CO_2 molecule identified by the blue arrow.

In the following, we further discuss the enhanced adsorption of CO_2 on WSSe nanotubes with the introduction of Se vacancies, from the aspects of CDD, electron transfer, and DOS. As mentioned before, after CO_2 adsorption at the Se vacancy site, C and O2 atoms separately bond to W atoms. As plotted in Figure 3b, the electron transfer amount from the defective Janus WSSe nanotube to the CO_2 molecules is up to 1.12 e. The formation of C-W and O-W bonds indicate that on the defective Janus WSSe nanotube, the CO_2 adsorption is chemical adsorption.

For the purpose of understanding the electronic origin of the chemisorption on the defective Janus WSSe nanotube, its corresponding DOS is calculated. For the defective Janus WSSe nanotube, its conduction band maximum (CBM) rises to a high level after the CO_2 adsorption (see Figure 4a,b), which corresponds to the Bader charge result that the defective Janus WSSe nanotube loses 1.12 e. In addition, as shown in Figure 4c, there is an obvious orbital hybridization between the CO_2 molecule and the defective Janus WSSe nanotube, which is mainly contributed by the O-p and C-p orbitals from the adsorbed molecule as well as the W-d orbitals from the W atoms in substrate bonding to the C and O2 atoms. This explains the phenomenon that the CO_2 gas molecule is tightly attached to the defective Janus WSSe nanotube through the C-W and O-W bonds. In addition, the DOS of the CO_2 molecules pre- and post-adsorption (see Figures 2a and S5) shows that an obvious delocalization of DOS occurs after adsorption, which means a severe electron redistribution in the adsorbed CO_2 gas molecule, caused by the gained electrons from the substrate. The results above provide more evidence that the adsorption of CO_2 by the defective WSSe nanotubes is chemisorption. In other words, the introduction of Se vacancy can well convert the physical adsorption of CO_2 into chemical adsorption on the Janus WSSe nanotube.

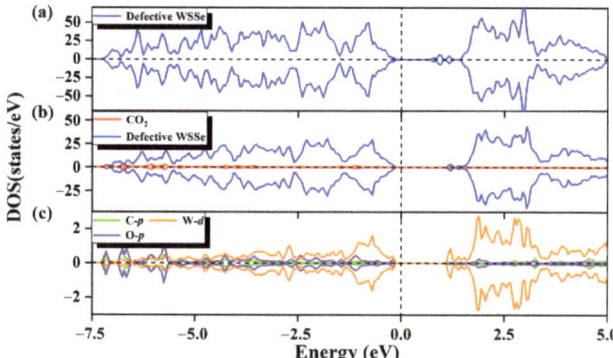

Figure 4. The total state density of WSSe nanotube with Se vacancy (**a**). The partial state density (**b**) of the adsorption system, the defective WSSe nanotubes are shown in dark blue, and CO_2 is shown in red. (**c**) Partial state densities of adsorbed CO_2 gas molecules in C p orbitals (cyan), O p orbitals (purple), and W d orbitals of two W atoms in substrate (orange). The vertical dashed line shows the Fermi level.

2.3. Photocatalytic Performance of Defective WSSe Nanotube for CO$_2$RR

The activation of the CO$_2$ gas molecule on the defective WSSe nanotube makes the further catalytic CO$_2$ reduction reaction possible. As displayed in Figure 5a, though the Se vacancy bring about some gap states, the defective Janus WSSe nanotube still keeps the semiconductor character with a narrower band gap of 0.83 eV (the band gap of the pristine Janus WSSe nanotube is 1.56 eV, see Figure 5b). In the following, we studied the photocatalytic performance of the defective Janus WSSe nanotube.

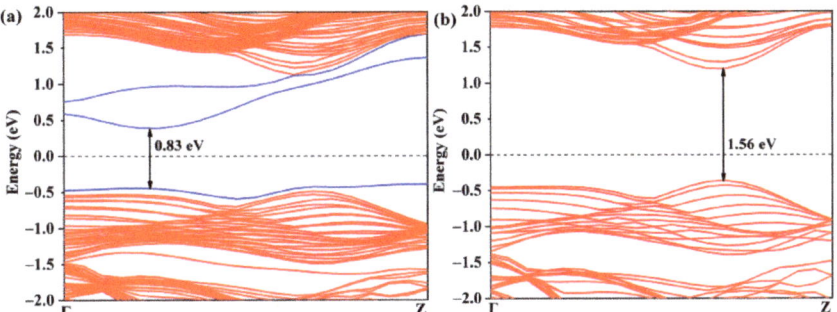

Figure 5. Band structures of (**a**) defective and (**b**) pristine Janus WSSe nanotubes. The gap states caused by the Se vacancy are marked with blue lines. The black number represents the value of band gap.

In order to initiate the photocatalytic conversion of CO$_2$, an efficient photocatalyst must have a high photo-conversion efficiency. As shown in Figure 6a, there are several significant light absorption peaks (over 10^5 cm^{-1}) among the visible light area for the pristine and defective Janus WSSe nanotubes, indicating they are promising catalyst candidates with visible-light responses. The highest absorption peak in the visible area for the pristine and defective Janus WSSe nanotubes arrive 3.10×10^5 cm^{-1} (at 380.00 nm, black line) and 2.96×10^5 cm^{-1} (at 380.00 nm, red line), which exceed the one of the planar Janus WSSe (1.30×10^5 cm^{-1} at 466.28 nm) [34] and are on par with some reported photocatalysts, namely, MoSSe/graphene (4.00×10^5 cm^{-1} at 500 nm) [36] and MoSSe/AlN (3.95×10^5 cm^{-1} at 412 nm) [37]. Although the difference between the light absorption spectra of the pristine and defective Janus WSSe nanotubes are not significant, as displayed in Figure S6, in the infrared and visible regions, the optical absorption coefficient of the defective Janus WSSe nanotube is higher than the one of the pristine Janus WSSe nanotube, which is consistent with the fact that the defective Janus WSSe nanotube has a smaller band gap than the pristine one. The non-zero absorbance value in the infrared region (IR) of the defective WSSe nanotube ensures the utilization of IR photons. Therefore, the introduction of Se vacancy defects makes the Janus WSSe nanotube use photons in a relatively larger energy range. Additionally, the negligible difference of light absorption spectra between these two kinds of nanotubes may be caused by the fact that the gap states are too weak in the defective Janus WSSe, where the concentration of Se vacancy is too low (just 4.17%). In the visible region, the reported optical absorption coefficient of defective Janus WSSe monolayer with a higher concentration of Se vacancy (6.25%) is more obviously higher than the pristine Janus WSSe monolayer [34], which agrees well with the results of nanotubes.

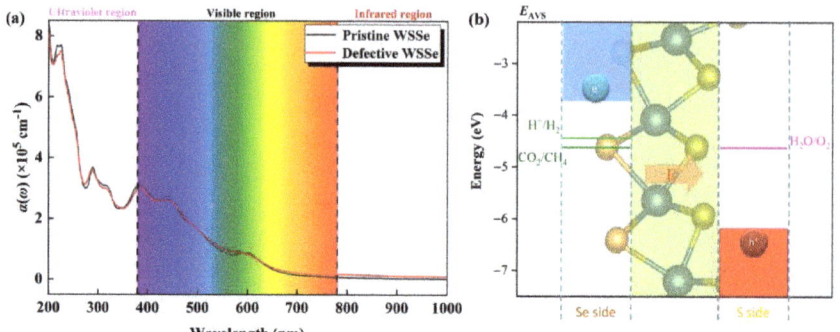

Figure 6. (a) Optical absorbance of pristine and defective Janus WSSe nanotubes. (b) Schematic diagram of band edge position of Janus WSSe nanotube relative to normal hydrogen electrode (NHE) at pH = 0. E_{AVS} represents the energy level relative to the absolute vacuum scale (AVS). The pink arrow represents the orientation of the built-in electric field.

In order for a semiconductor to be active for photo-reduction of CO_2, the band edges must be aligned with the potentials of the reduction half-reactions [38]. On top of that, its band edge also needs to satisfy the oxidation potential of H_2O/O_2 because the oxygen evolution reaction (OER) could consume the redundant photo-excited holes and provide the necessary $H^+ + e^-$ pair simultaneously. As shown in Figure 6b, the CBM in the photocatalytic redox capacity is above the CO_2/CH_4 reduction potential, and the VBM is below the H_2O/O_2 oxidation potential, indicating that the WSSe nanotubes have sufficient redox capacity for both photocatalytic CO_2RR and OER. Furthermore, our previous work pointed out that [27] the dipole caused by the structural asymmetry introduces a built-in electric field with the direction from the Se layer to the S layer (see the pink arrow in Figure 6b). In this case, the photoexcited electron and hole will run fast in opposite directions, causing high spatial separation of the electron–hole pairs, which surely suppresses the recombination of photoexcited carriers.

Next, we explore whether the reaction can be spontaneous under dynamic conditions. The case without any external potential (U = 0 V) is used to simulate the condition in darkness. We first screen the favorable reaction path of CO_2RR on the defective Janus WSSe nanotube (see Figure S7). The CO_2RR-to-CH_4 process involves eight proton-coupled electron transfer steps ($CO_2 + 8H^+ + 8e^- \rightarrow 2H_2O + CH_4$). The free energy diagram and the corresponding intermediates for the CO_2RR-to-CH_4 are shown in Figure 7a. The most possible path is $CO_2 * \rightarrow OCOH * \rightarrow OCHOH * \rightarrow OCH * \rightarrow OCH_2 * \rightarrow OCH_3 * \rightarrow O * \rightarrow OH * \rightarrow H_2O *$. The electrocatalytic steps, i.e., OCHOH * $\rightarrow$ OCH *, OCH * $\rightarrow$ OCH_2 *, OCH_2 * $\rightarrow$ OCH_3 *, and OCH_3 * $\rightarrow$ O *, are exothermic by −0.41, −0.51, −0.15, and −1.43 eV, respectively; meanwhile, the other hydrogenation steps, i.e., $CO_2 * \rightarrow OCOH *$, OCOH * $\rightarrow$ OCHOH *, O * $\rightarrow$ OH *, and OH * $\rightarrow$ $H_2O *$, are endothermic by 0.65, 0.15, 0.06, and 0.50 eV, respectively. The formation of OCOH * is the potential determining step (PDS) with a limiting potential (U_l) of −0.65 V. At the same time, we also investigated the OER process on the S side of the defective Janus WSSe nanotube along the 4 e transfer pathway, i.e., $H_2O \rightarrow OH * \rightarrow OOH * \rightarrow O_2$ (see Figure 7b) [18,27]. The free energy changes (ΔG) for the four different steps are endothermic by 1.81, 0.06, 1.75, and 1.30 eV, respectively. The formation of OH * is the PDS with a U_l of −1.81 V.

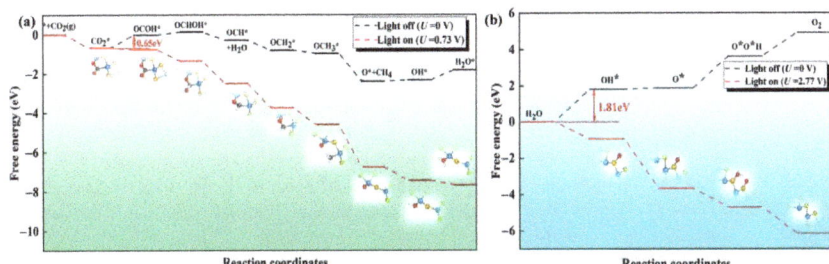

Figure 7. The Gibbs free energy diagrams for the (**a**) 8 e pathway of CO_2RR and (**b**) 4 e pathway of OER on the defective Janus WSSe nanotube under different light conditions. The extra potentials provided by photogenerated electrons and holes are 0.73 and 2.77 V, respectively.

According to the free energy calculations mentioned above, it could be found that both the CO_2RR and OER have endothermic steps; thus, they could not take place spontaneously without photo-irradiation. However, the high enough external potential supplied by the photo-excited carriers helps to overcome the U_l of these redox half-reactions, making the redox half-reactions proceed spontaneously [39]. The extra potential of the photogenerated electrons/holes (U_e/U_h) is defined as the energy difference between H^+/H_2 reduction potential and the CBM/VBM [18,39–41]. According to our previous work [27], the U_e and U_h of the defective Janus WSSe nanotube at pH = 0 are 0.73 and 2.77 V, respectively, which are sufficient enough to separately cover the U_l of CO_2RR and OER. Therefore, in consideration of U_e and U_h, all the reduction and oxidation steps become downhill (red dash lines in Figure 7a,b). That is to say, under the light irradiation, both CO_2RR and OER can operate spontaneously.

Usually, the hydrogen evolution reaction (HER) is considered to be an important competitive side reaction in the catalytic CO_2RR [42,43]. Next, we investigated the competitive relationship between CO_2RR and HER in the defective Janus WSSe nanotubes. Based on the Brønsted-Evans-Polanyi relation [44,45], the reaction with lower Gibbs, ΔG, values has a smaller reaction barrier; thus, it is more favorable for kinetics. Accordingly, the ΔG for H * formation energy (ΔG_{H^*}) is calculated (Figure 8a) and compared with the one for CO_2 * formation energy (ΔG_{CO_2*}). As shown in Figure 8b, ΔG_{CO_2*} (−0.67 eV) is more negative than ΔG_{H^*} (−0.15 eV), which ensures that the active sites are preferred to be occupied by CO_2 *. Therefore, the defective Janus WSSe nanotube is more selective for CO_2RR over HER.

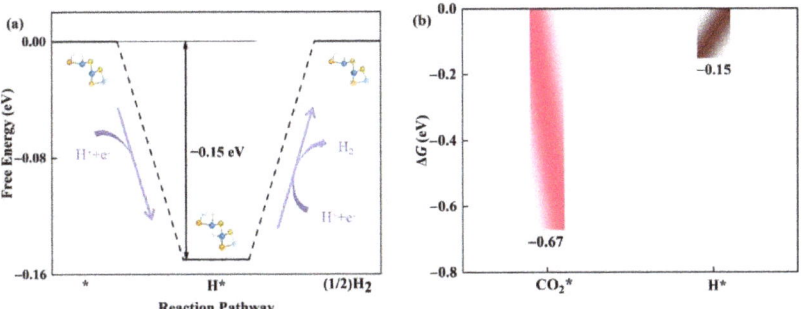

Figure 8. (**a**) Gibbs free energy diagram of HER on defective Janus WSSe nanotube. (**b**) ΔG_{CO_2*} (pink bar) vs. ΔG_{H^*} (brown bar) of defective Janus WSSe nanotube. * means the adsorption site.

3. Computational Methods

In our work, all the computational models are constructed with the DeviceStudio software [46]. In addition, the Geometric relaxation and electronic structure were conducted based on DFT simulations employing DS-PAW software [47]. The exchange-correlation energy of Perdew–Burke–Ernzerhof (PBE) was employed [48]. To depict the van der Waals (vdW) coupling in the adsorption system, we used the zero-damping DFT-D3 method suggested from Grimme [49]. All internal coordinates with fixed lattice constants were permitted to relax during the optimization process. The sampling integration of the Brillouin zone was performed in accordance with the Monkhorst-Pack scheme [50], and the structure optimization and electronic properties are calculated with a $1 \times 1 \times 4$ K-point. The value of 500 eV was chosen as the cutoff energy of plane-wave basis. We set the periodic boundary condition along the z-axis and put more than 10.8 Å vacuum spaces along the x and y directions to avoid the interaction between adjacent nanotubes. Periodic boundary conditions were set on the z-axis and a vacuum space of more than 10.8 Å was applied on the x- and y-axes to evade adjacent nanotubes from interacting with each other. Moreover, the ΔG of CO_2RR and OER were calculated using the computational hydrogen electrode (CHE) model [51]. Additional details of the Gibbs free energy simulations are available in the Supporting Information.

The E_{ads} of the CO_2 on the WSSe nanotube was obtained from the following equation [52,53],

$$E_{ads} = E_{total} - E_{sub} - E_{CO_2} \tag{1}$$

where E_{total} was the total energy of the adsorption system, while E_{CO_2} and E_{sub} separately were the total energies of the isolated CO_2 molecule and the clean Janus WSSe nanotube. A higher negative E_{ads} indicated a more favorable exothermic adsorption.

The plane-integrated CDD was carried out in accordance with the equation,

$$\Delta \rho = \rho_{total} - \rho_{sub} - \rho_{CO_2} \tag{2}$$

where ρ_{total}, ρ_{CO_2}, and ρ_{sub} separately were the charge density of the adsorption system, adsorbed CO_2 molecule, and substrate.

The absorption coefficient, $a(\omega)$, used to estimate the solar energy gathering capacity was given by the following equation [27],

$$a(\omega) = \sqrt{2} \frac{\omega}{c} \left(\sqrt{\varepsilon_1(\omega)^2 + \varepsilon_2(\omega)^2} - \varepsilon_1(\omega) \right)^{\frac{1}{2}} \tag{3}$$

where ε_1 and ε_2 frequently were the real and imaginary parts of the frequency-dependent dielectric function, while c was the speed of light under vacuum.

4. Conclusions

In this paper, based on the first-principles calculations, we investigate the performance of the defective Janus WSSe nanotube for the photocatalytic CO_2RR. The introduction of Se vacancy could significantly increase the amount of interfacial transferred electrons and lead to obvious electron orbital hybridization between adsorbates and substrates, making the CO_2 adsorption on the Janus WSSe nanotube transform into chemisorption from physisorption. Strong chemisorption enables defective Janus WSSe nanotubes to be highly active and selective against CO_2RR. In addition, the extra potential from photo-produced carriers is high enough to trigger spontaneous CO_2RR and OER simultaneously on the defective Janus WSSe nanotube. For the first time, our work theoretically predicts the high photocatalytic performance of the defective Janus WSSe nanotube on CO_2RR, which promisingly will stimulate extensive interests from material science and chemistry communities to realize our vision.

Supplementary Materials: The following Supporting information can be downloaded at: https://www.mdpi.com/article/10.3390/molecules28124602/s1, Figure S1: Diameter, W-S bond length, W-S bond length and Se-S height of the pristine Janus WSSe nanotube; Figure S2: The relative adsorption energy and the location for the CO_2 gas molecule passing through the pristine WSSe nanotube wall; Figure S3: Relative value of total energy variations as well as their corresponding fittings for the pristine and defective Janus WSSe nanotubes with respect to Strain ε along the tube axis; Figure S4: Top view and Side view of CO_2 gas molecules adsorbed at the Se vacancy of Janus WSSe monolayer; Figure S5: The enlarged view for the partial density of States of CO_2 portion from the adsorption System; Figure S6: The enlarged view the optical absorbance of pristine and defective Janus WSSe nanotubes at wavelength of 500–900 nm; Figure S7: The Search process for the minimum energy reaction pathways of the CO_2 reduction reactions on defective Janus WSSe nanotube; Table S1: The amount of charge transfer for C and O atoms of CO_2 gas molecules adsorbed in pristine and defective Janus WSSe nanotubes, respectively; Free energy difference in the CO2RR and OER; Table S2: Zero-pint energy correction and entropy contribution of molecules and adsorbates in this Study. Refs. [51,54] are cited in Supplementary Materials.

Author Contributions: Software, H.D. and S.Y.; Validation, J.L. and Y.G.; Formal analysis, L.J. and X.T.; Investigation, X.T., J.L., Y.G. and W.L.; Resources, L.J.; Data curation, X.T. and H.D.; Writing—original draft, L.J., X.T. and J.L.; Supervision, L.J.; Funding acquisition, L.J. All authors have read and agreed to the published version of the manuscript.

Funding: Our work is funded by the Natural Science Foundation of Henan Province (Grant No. 232300420128), National College Students Innovation and Entrepreneurship Training Program (Grant No. 202210479032), Open Project of Key Laboratory of Functional Materials and Devices for Informatics of Anhui Higher Education Institutes (Grant No. FSKFKT002), College Students Innovation Fund of Anyang Normal University (Grant No. 202210479049), and Key Scientific and Technological Projects in Anyang City (Grant No. 2022C01GX019).

Institutional Review Board Statement: Not applicable.

Informed Consent Statement: Not applicable.

Data Availability Statement: The data presented in this study are available in Supplementary Materials.

Conflicts of Interest: The authors declare no conflict of interest.

Sample Availability: Not applicable.

References

1. Gattrell, M.; Gupta, N.; Co, A. A review of the aqueous electrochemical reduction of CO_2 to hydrocarbons at copper. *J. Electroanal. Chem.* **2006**, *594*, 1–19. [CrossRef]
2. Gattrell, M.; Gupta, N.; Co, A. Electrochemical reduction of CO_2 to hydrocarbons to store renewable electrical energy and upgrade biogas. *Energy Convers. Manag.* **2007**, *48*, 1255–1265. [CrossRef]
3. Wageh, S.; Al-Hartomy, O.A.; Alotaibi, M.F.; Liu, L.-J. Ionized cocatalyst to promote CO_2 photoreduction activity over core–triple-shell ZnO hollow spheres. *Rare Met.* **2022**, *41*, 1077–1079. [CrossRef]
4. Chen, Y.; Sun, Q.; Jena, P. SiTe monolayers: Si-based analogues of phosphorene. *J. Mater. Chem. C* **2016**, *4*, 6353–6361. [CrossRef]
5. Modestra, J.A.; Mohan, S.V. Microbial electrosynthesis of carboxylic acids through CO_2 reduction with selectively enriched biocatalyst: Microbial dynamics. *J. CO_2 Util.* **2017**, *20*, 190–199. [CrossRef]
6. Cai, F.; Gao, D.; Zhou, H.; Wang, G.; He, T.; Gong, H.; Miao, S.; Yang, F.; Wang, J.; Bao, X. Electrochemical promotion of catalysis over Pd nanoparticles for CO_2 reduction. *Chem. Sci.* **2017**, *8*, 2569–2573. [CrossRef]
7. Wang, Y.-H.; Jiang, W.-J.; Yao, W.; Liu, Z.-L.; Liu, Z.; Yang, Y.; Gao, L.-Z. Advances in electrochemical reduction of carbon dioxide to formate over bismuth-based catalysts. *Rare Met.* **2021**, *40*, 2327–2353. [CrossRef]
8. Qiao, L.; Song, M.; Geng, A.; Yao, S. Polyoxometalate-based high-nuclear cobalt–vanadium–oxo cluster as efficient catalyst for visible light-driven CO2 reduction. *Chin. Chem. Lett.* **2019**, *30*, 1273–1276. [CrossRef]
9. Wang, Y.; Tian, Y.; Yan, L.; Su, Z. DFT study on sulfur-doped g-C_3N_4 nanosheets as a photocatalyst for CO_2 reduction reaction. *J. Phys. Chem. C* **2018**, *122*, 7712–7719. [CrossRef]
10. Tackett, B.M.; Gomez, E.; Chen, J.G. Net reduction of CO_2 via its thermocatalytic and electrocatalytic transformation reactions in standard and hybrid processes. *Nat. Catal.* **2019**, *2*, 381–386. [CrossRef]
11. Wang, X.-T.; Lin, X.-F.; Yu, D.-S. Metal-containing covalent organic framework: A new type of photo/electrocatalyst. *Rare Met.* **2021**, *41*, 1160–1175. [CrossRef]

12. Zhou, A.-Q.; Yang, J.-M.; Zhu, X.-W.; Zhu, X.-L.; Liu, J.-Y.; Zhong, K.; Chen, H.-X.; Chu, J.-Y.; Du, Y.-S.; Song, Y.-H.; et al. Self-assembly construction of NiCo LDH/ultrathin g-C3N4 nanosheets photocatalyst for enhanced CO_2 reduction and charge separation mechanism study. *Rare Met.* **2022**, *41*, 2118–2128. [CrossRef]
13. Muiruri, J.K.; Ye, E.; Zhu, Q.; Loh, X.J.; Li, Z. Recent advance in nanostructured materials innovation towards photocatalytic CO_2 reduction. *Appl. Catal. A Gen.* **2022**, *648*, 118927. [CrossRef]
14. Luo, Z.; Li, Y.; Guo, F.; Zhang, K.; Liu, K.; Jia, W.; Zhao, Y.; Sun, Y. Carbon Dioxide Conversion with High-Performance Photocatalysis into Methanol on NiSe2/WSe2. *Energies* **2020**, *13*, 4330. [CrossRef]
15. Biswas, M.; Ali, A.; Cho, K.Y.; Oh, W.C. Novel synthesis of WSe2-Graphene-TiO2 ternary nanocomposite via ultrasonic technics for high photocatalytic reduction of CO_2 into CH_3OH. *Ultrason. Sonochem.* **2018**, *42*, 738–746. [CrossRef]
16. Ali, A.; Oh, W.C. Preparation of Nanowire like WSe2-Graphene Nanocomposite for Photocatalytic Reduction of CO_2 into CH_3OH with the Presence of Sacrificial Agents. *Sci. Rep.* **2017**, *7*, 1867. [CrossRef]
17. Ju, L.; Bie, M.; Shang, J.; Tang, X.; Kou, L. Janus transition metal dichalcogenides: A superior platform for photocatalytic water splitting. *J. Phys. Mater.* **2020**, *3*, 22004. [CrossRef]
18. Ju, L.; Shang, J.; Tang, X.; Kou, L. Tunable Photocatalytic Water Splitting by the Ferroelectric Switch in a 2D AgBiP2Se6 Monolayer. *J. Am. Chem. Soc.* **2020**, *142*, 1492–1500. [CrossRef]
19. Ju, L.; Bie, M.; Tang, X.; Shang, J.; Kou, L. Janus WSSe Monolayer: An Excellent Photocatalyst for Overall Water Splitting. *ACS Appl. Mater. Interfaces* **2020**, *12*, 29335–29343. [CrossRef]
20. Lingampalli, S.R.; Ayyub, M.M.; Rao, C.N.R. Recent Progress in the Photocatalytic Reduction of Carbon Dioxide. *ACS Omega* **2017**, *2*, 2740–2748. [CrossRef]
21. Li, X.; Wen, J.; Low, J.; Fang, Y.; Yu, J. Design and fabrication of semiconductor photocatalyst for photocatalytic reduction of CO_2 to solar fuel. *Sci. China Mater.* **2014**, *57*, 70–100. [CrossRef]
22. Tu, W.; Zhou, Y.; Zou, Z. Photocatalytic conversion of CO_2 into renewable hydrocarbon fuels: State-of-the-art accomplishment, challenges, and prospects. *Adv. Mater.* **2014**, *26*, 4607–4626. [CrossRef] [PubMed]
23. Zhang, J.; Jia, S.; Kholmanov, I.; Dong, L.; Er, D.; Chen, W.; Guo, H.; Jin, Z.; Shenoy, V.B.; Shi, L. Janus monolayer transition-metal dichalcogenides. *ACS Nano* **2017**, *11*, 8192–8198. [CrossRef] [PubMed]
24. Lu, A.-Y.; Zhu, H.; Xiao, J.; Chuu, C.-P.; Han, Y.; Chiu, M.-H.; Cheng, C.-C.; Yang, C.-W.; Wei, K.-H.; Yang, Y. Janus monolayers of transition metal dichalcogenides. *Nat. Nanotechnol.* **2017**, *12*, 744–749. [CrossRef]
25. Ma, X.; Wu, X.; Wang, H.; Wang, Y. A Janus MoSSe monolayer: A potential wide solar-spectrum water-splitting photocatalyst with a low carrier recombination rate. *J. Mater. Chem. A* **2018**, *6*, 2295–2301. [CrossRef]
26. Xia, C.; Xiong, W.; Du, J.; Wang, T.; Peng, Y.; Li, J. Universality of electronic characteristics and photocatalyst applications in the two-dimensional Janus transition metal dichalcogenides. *Phys. Rev. B* **2018**, *98*, 165424. [CrossRef]
27. Ju, L.; Liu, P.; Yang, Y.; Shi, L.; Yang, G.; Sun, L. Tuning the photocatalytic water-splitting performance with the adjustment of diameter in an armchair WSSe nanotube. *J. Energy Chem.* **2021**, *61*, 228–235. [CrossRef]
28. Evarestov, R.A.; Kovalenko, A.V.; Bandura, A.V. First-principles study on stability, structural and electronic properties of monolayers and nanotubes based on pure Mo(W)S(Se)2 and mixed (Janus) Mo(W)SSe dichalcogenides. *Phys. E Low-Dimens. Syst. Nanostruct.* **2020**, *115*, 113681. [CrossRef]
29. Ju, L.; Dai, Y.; Wei, W.; Li, M.; Huang, B. DFT investigation on two-dimensional GeS/WS2 van der Waals heterostructure for direct Z-scheme photocatalytic overall water splitting. *Appl. Surf. Sci.* **2018**, *434*, 365–374. [CrossRef]
30. Ju, L.; Liu, C.; Shi, L.; Sun, L. The high-speed channel made of metal for interfacial charge transfer in Z-scheme g–C_3N_4/MoS2 water-splitting photocatalyst. *Mater. Res. Express* **2019**, *6*, 115545. [CrossRef]
31. Lin, H.-F.; Liu, L.-M.; Zhao, J. 2D lateral heterostructures of monolayer and bilayer phosphorene. *J. Mater. Chem. C* **2017**, *5*, 2291–2300. [CrossRef]
32. Ma, D.; Ju, W.; Li, T.; Zhang, X.; He, C.; Ma, B.; Lu, Z.; Yang, Z. The adsorption of CO and NO on the MoS2 monolayer doped with Au, Pt, Pd, or Ni: A first-principles study. *Appl. Surf. Sci.* **2016**, *383*, 98–105. [CrossRef]
33. Ju, L.; Tang, X.; Li, X.; Liu, B.; Qiao, X.; Wang, Z.; Yin, H. NO_2 Physical-to-Chemical Adsorption Transition on Janus WSSe Monolayers Realized by Defect Introduction. *Molecules* **2023**, *28*, 1644. [CrossRef]
34. Ju, L.; Tang, X.; Zhang, Y.; Li, X.; Cui, X.; Yang, G. Single Selenium Atomic Vacancy Enabled Efficient Visible-Light-Response Photocatalytic NO Reduction to NH_3 on Janus WSSe Monolayer. *Molecules* **2023**, *28*, 2959. [CrossRef]
35. Zhang, J.; Tang, X.; Chen, M.; Ma, D.; Ju, L. Tunable Photocatalytic Water Splitting Performance of Armchair MoSSe Nanotubes Realized by Polarization Engineering. *Inorg. Chem.* **2022**, *61*, 17353–17361. [CrossRef]
36. Deng, S.; Li, L.; Rees, P. Graphene/MoXY Heterostructures Adjusted by Interlayer Distance, External Electric Field, and Strain for Tunable Devices. *ACS Appl. Nano Mater.* **2019**, *2*, 3977–3988. [CrossRef]
37. Ren, K.; Wang, S.; Luo, Y.; Chou, J.-P.; Yu, J.; Tang, W.; Sun, M. High-efficiency photocatalyst for water splitting: A Janus MoSSe/XN (X = Ga, Al) van der Waals heterostructure. *J. Phys. D Appl. Phys.* **2020**, *53*, 185504. [CrossRef]
38. Fan, Y.; Song, X.; Ai, H.; Li, W.; Zhao, M. Highly Efficient Photocatalytic CO_2 Reduction in Two-Dimensional Ferroelectric CuInP2S6 Bilayers. *ACS Appl Mater Interfaces* **2021**, *13*, 34486–34494. [CrossRef]
39. Qiao, M.; Liu, J.; Wang, Y.; Li, Y.; Chen, Z. PdSeO3 Monolayer: Promising Inorganic 2D Photocatalyst for Direct Overall Water Splitting Without Using Sacrificial Reagents and Cocatalysts. *J. Am. Chem. Soc.* **2018**, *140*, 12256–12262. [CrossRef]

40. Greeley, J.; Jaramillo, T.F.; Bonde, J.; Chorkendorff, I.; Nørskov, J.K. Computational high-throughput screening of electrocatalytic materials for hydrogen evolution. *Nat. Mater.* **2006**, *5*, 909–913. [CrossRef]
41. Rossmeisl, J.; Qu, Z.W.; Zhu, H.; Kroes, G.J.; Nørskov, J.K. Electrolysis of water on oxide surfaces. *J. Electroanal. Chem.* **2007**, *607*, 83–89. [CrossRef]
42. Goyal, A.; Marcandalli, G.; Mints, V.A.; Koper, M.T.M. Competition between CO_2 Reduction and Hydrogen Evolution on a Gold Electrode under Well-Defined Mass Transport Conditions. *J. Am. Chem. Soc.* **2020**, *142*, 4154–4161. [CrossRef] [PubMed]
43. Ooka, H.; Figueiredo, M.C.; Koper, M.T.M. Competition between Hydrogen Evolution and Carbon Dioxide Reduction on Copper Electrodes in Mildly Acidic Media. *Langmuir* **2017**, *33*, 9307–9313. [CrossRef] [PubMed]
44. Bronsted, J. Acid and Basic Catalysis. *Chem. Rev.* **1928**, *5*, 231–338. [CrossRef]
45. Evans, M.; Polanyi, M. Inertia and driving force of chemical reactions. *Trans. Faraday Soc.* **1938**, *34*, 11–24. [CrossRef]
46. Hongzhiwei Technology, D.S., Version 2022B, China. 2022. Available online: https://iresearch.net.cn/cloudSoftware (accessed on 14 December 2022).
47. Blöchl, P.E. Projector augmented-wave method. *Phys. Rev. B* **1994**, *50*, 17953–17979. [CrossRef]
48. Perdew, J.P.; Burke, K.; Ernzerhof, M. Generalized Gradient Approximation Made Simple. *Phys. Rev. Lett.* **1996**, *77*, 3865–3868. [CrossRef]
49. Grimme, S.; Antony, J.; Ehrlich, S.; Krieg, H. A consistent and accurate ab initio parametrization of density functional dispersion correction (DFT-D) for the 94 elements H-Pu. *J. Chem. Phys.* **2010**, *132*, 154104. [CrossRef]
50. Monkhorst, H.J.; Pack, J.D. Special points for Brillouin-zone integrations. *Phys. Rev. B* **1976**, *13*, 5188. [CrossRef]
51. Nørskov, J.K.; Rossmeisl, J.; Logadottir, A.; Lindqvist, L.; Kitchin, J.R.; Bligaard, T.; Jonsson, H. Origin of the overpotential for oxygen reduction at a fuel-cell cathode. *J. Phys. Chem. B* **2004**, *108*, 17886–17892. [CrossRef]
52. Li, D.-H.; Li, Q.-M.; Qi, S.-L.; Qin, H.-C.; Liang, X.-Q.; Li, L. Theoretical Study of Hydrogen Production from Ammonia Borane Catalyzed by Metal and Non-Metal Diatom-Doped Cobalt Phosphide. *Molecules* **2022**, *27*, 8206. [CrossRef]
53. Liu, X.; Xu, Y.; Sheng, L. Al-Decorated C_2N Monolayer as a Potential Catalyst for NO Reduction with CO Molecules: A DFT Investigation. *Molecules* **2022**, *27*, 5790. [CrossRef]
54. Li, X.; Dai, Y.; Ma, Y.; Li, M.; Yu, L.; Huang, B. Landscape of DNA-like inorganic metal free double helical semiconductors and potential applications in photocatalytic water splitting. *J. Mater. Chem. A* **2017**, *5*, 8484–8492. [CrossRef]

Disclaimer/Publisher's Note: The statements, opinions and data contained in all publications are solely those of the individual author(s) and contributor(s) and not of MDPI and/or the editor(s). MDPI and/or the editor(s) disclaim responsibility for any injury to people or property resulting from any ideas, methods, instructions or products referred to in the content.

Article

Achieving Boron–Carbon–Nitrogen Heterostructures by Collision Fusion of Carbon Nanotubes and Boron Nitride Nanotubes

Chao Zhang [1,*], Jiangwei Xu [1], Huaizhi Song [1], Kai Ren [2], Zhi Gen Yu [3] and Yong-Wei Zhang [3]

[1] School of Materials Science and Engineering, Anhui University of Science and Technology, Huainan 232001, China; xujw1022@163.com (J.X.); huaizhi.song@outlook.com (H.S.)
[2] School of Mechanical and Electronic Engineering, Nanjing Forestry University, Nanjing 210042, China; kairen@njfu.edu.cn
[3] Institute of High Performance Computing (IHPC), Agency for Science, Technology and Research (A*STAR), 1 Fusionopolis Way, #16-16 Connexis, Singapore 138632, Singapore; yuzg@ihpc.a-star.edu.sg (Z.G.Y.); zhangyw@ihpc.a-star.edu.sg (Y.-W.Z.)
* Correspondence: chaozhang@mail.bnu.edu.cn

Citation: Zhang, C.; Xu, J.; Song, H.; Ren, K.; Yu, Z.G.; Zhang, Y.-W. Achieving Boron–Carbon–Nitrogen Heterostructures by Collision Fusion of Carbon Nanotubes and Boron Nitride Nanotubes. *Molecules* **2023**, *28*, 4334. https://doi.org/10.3390/molecules28114334

Academic Editor: Lin Ju

Received: 20 April 2023
Revised: 22 May 2023
Accepted: 24 May 2023
Published: 25 May 2023

Copyright: © 2023 by the authors. Licensee MDPI, Basel, Switzerland. This article is an open access article distributed under the terms and conditions of the Creative Commons Attribution (CC BY) license (https://creativecommons.org/licenses/by/4.0/).

Abstract: Heterostructures may exhibit completely new physical properties that may be otherwise absent in their individual component materials. However, how to precisely grow or assemble desired complex heterostructures is still a significant challenge. In this work, the collision dynamics of a carbon nanotube and a boron nitride nanotube under different collision modes were investigated using the self-consistent-charge density-functional tight-binding molecular dynamics method. The energetic stability and electronic structures of the heterostructure after collision were calculated using the first-principles calculations. Five main collision outcomes are observed, that is, two nanotubes can (1) bounce back, (2) connect, (3) fuse into a defect-free BCN heteronanotube with a larger diameter, (4) form a heteronanoribbon of graphene and hexagonal boron nitride and (5) create serious damage after collision. It was found that both the BCN single-wall nanotube and the heteronanoribbon created by collision are the direct band-gap semiconductors with the band gaps of 0.808 eV and 0.544 eV, respectively. These results indicate that collision fusion is a viable method to create various complex heterostructures with new physical properties.

Keywords: heteronanotube; heteronanoribbon; collision dynamics; electronic properties; DFTB

1. Introduction

In recent decades, research interest in nanomaterials' synthesis has been growing rapidly, enabling the attainment of novel materials such as heterostructures [1–6]. These heterostructures often exhibit fascinating new physical phenomena and properties that are otherwise absent in their individual components [7–11]. Currently, it is still a significant challenge to precisely grow or assemble such heterostructures [12–16].

Carbon nanotubes (CNTs) [17], one of the most widely studied nanomaterials, can be regarded as one-dimensional nanomaterials with unique hollow structures formed by curling up graphene sheets. They can exhibit excellent electrical, thermal, mechanical and optical performances, and have been widely used in electronic devices, hydrogen storage, thermally conductive materials and composite materials [18–21]. Similar to CNTs, boron nitride nanotubes (BNNTs) can also be regarded as hollow tubes formed by rolling up hexagonal boron nitride (*h*-BN) sheets [22–26], which have a hexagonal layered structure similar to graphene. BNNTs are wide-band-gap semiconductors with high thermal conductivity, high oxidation resistance and excellent mechanical strength, and they have been widely used in the field of nanocomposites and hydrogen storage materials [27–29]. Note that BNNTs have a similar structure to CNTs and their difference is that only nitrogen and boron atoms are alternately replaced by carbon atoms. The lattice mismatch between BNNT and CNT is about 0.79%. This tiny lattice difference provides a possibility of designing heterostructural materials by combining these two nanostructures. At present, there are generally two ways to form heteronanotubes by combining BNNTs and CNTs. One way is to stack them together to form van der Waals co-axial bilayer heteronanotubes with CNT@BNNT, that is, single-walled BNNT-wrapped single-walled CNT (SWCNT) [30,31]. The CNT@BNNTs exhibit superior thermal stability, chemical inertness and mechanical robustness in comparison to CNTs [31]. Another method is to combine CNTs and BNNTs to form covalent heteronanotubes either along or perpendicular to the tube axis. The covalent heteronanotubes along the tube axis, which exhibit tunable electronic properties based on stoichiometry, were prepared by covalent doping of CNTs with BNNTs [31,32]. However, many complex covalent heteronanotubes, for example, perpendicular to the tube axis, have not been realized thus far. How to form various complex heteronanotubes remains largely unexplored, which motivates the present study. It is noted that doping into graphite or carbon nanotubes has been used to create BCN nanostructures, and the preparation method can be traced back to the work of Stephan et al. in 1994 [32]. Subsequently, $B_xC_yN_z$, CB_xN_y, CB_x and CN_x nanotubes were also successfully prepared. A large number of studies [33,34] have demonstrated that the doping of boron and nitrogen into carbon nanotubes can be an effective way to adjust the band gap of the material [35]. Interestingly, Campbell et al. obtained direct experimental evidence of the collision fusion of two C_{60} molecules [36–38].

In this work, BCN nanotubes were achieved by colliding CNTs and BNNTs based on the computational simulation method. To demonstrate the viability of this method, the self-consistent-charge density-functional tight-binding (SCC-DFTB) molecular dynamics (MD) method was employed to simulate the collision process. The simulations clearly demonstrated the formation of BCN nanotubes, indicating the parallel collision method is viable for producing BCN nanotubes. Interestingly, heteronanoribbons of graphene and h-BN are also formed after collision. We further showed that the BCN nanotube and the heteronanoribbon formed by the collision are direct band-gap semiconductors with band gaps of 0.81 eV and 1.34 eV, respectively. The present study suggests that heteronanotubes and heteronanoribbons can be effectively achieved by the collision fusion of CNTs and BNNTs.

2. Collision Results

There are five main collision outcomes that are observed in the simulations. I. The two nanotubes bounce off after collision and then move in opposite directions, labeled as B (see Video S1). II. The two nanotubes are merged together in the form of sp^3 hybridization after collision, labeled as C (see Figure 1a and Video S2). III. The two nanotubes are fused into a perfect (12,0) BCN single-walled nanotube with a larger diameter without any defects, labeled as P (see Figure 1b and Video S3). IV. The two nanotubes are either fused to form a larger diameter BCN nanotube with defects, including a 4–8 ring defect, 5–7 pair defect [39], Stone–Wales (SW) defect [40], inverse Stone–Wales (ISW) defect [41], etc. (see Figure 1c), or form a quasi-one-dimensional BCN nanoribbon with six-membered rings at the junction and irregular defects at the edge (see Figure 1d and Video S4), labeled as D. V. The two nanotubes are severely damaged after collision. With the increase in colliding energy, the structural damage becomes more and more serious, and the nanotubes may break up into single atoms, dimers, trimers or atomic chains, labeled as S.

The probability of each collision outcome is shown in Figure 2. It can be seen that the probability of the two nanotubes being bounced back and forming a heterojunction is higher in the lower energy range (0.1~0.3 eV). In the medium-energy region (0.3~0.5 eV), the probability of the two tubes fusing to form nanotubes and nanoribbons is relatively high. In the high-energy region (0.5~1.0 eV), the probability of fused structures with severe damage is very high. The simulation results are summarized as shown in Table 1. It can be observed that the collision mode has a minor influence on the collision results. The connected heterojunctions mainly occur in the collision mode of AA, h-AH and AB, and the corresponding energy is in the range of 0.1 to 0.3 eV. The fusion of two nanotubes into a perfect defect-free BCN nanotube mainly occurs in three collision modes, i.e., h-AH, AH and AB. The fused BCN nanotubes and BCN nanoribbons with defective structures are found in each collision mode, and the corresponding energy is in the range of 0.3 to 0.7 eV.

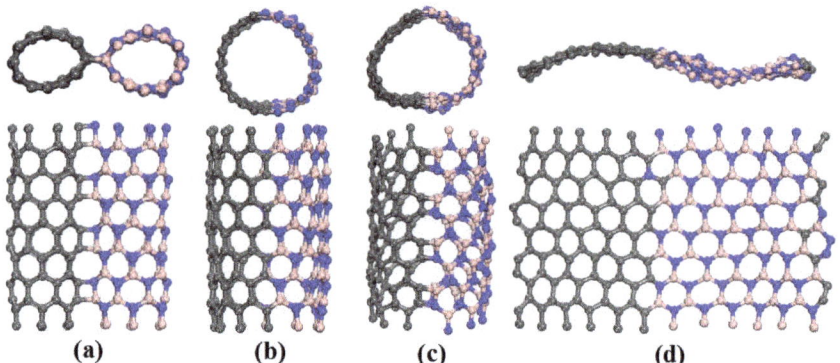

Figure 1. Typical structures formed by the collisions of CNTs and BNNTs. (**a**) Heterojunction, (**b**) defect-free (12,0) BCN nanotube, (**c**) BCN nanotube with four- and eight-membered ring defects, and (**d**) BCN nanoribbon.

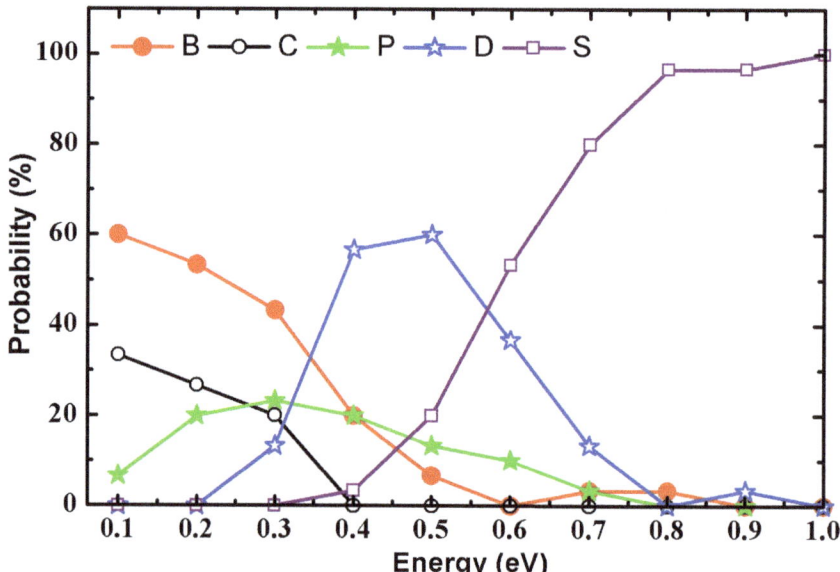

Figure 2. Probability of collision outcomes as a function of the initial collision energy. Solid circle, hollow circle, solid pentastar, hollow pentastar and hollow quadrangle represent the two nanotubes bouncing back (B) and connecting (C), fusing into BCN heteronanotubes without defects (P) and with defects (D) and colliding to create serious damage (S), respectively.

Table 1. Results of collision between SWCNT and SWBNNT. Table legend: B = bounce back, C = connecting heterojunction, P = perfect BCN nanotube with no defects, D = BCN nanotube with defects, S = serious damage, En = Energy.

Result	Model				
En (eV)	AA	h-AH	AH	AAH	AB
0.1	B, C, P	B, C, P	B, C	B	B, C
0.2	B, C	B, C, P	B, P	B	B, C, P
0.3	B, C	B, P, D	B, P, D	B, C, P, D	B, C, P
0.4	B, P, D	P, D	B, D	D	B, P, D
0.5	D	D, S	B, P, D	B, D, S	P, D, S
0.6	D, S	D, S	P, D, S	D, S	P, D, S
0.7	S	S	B, D, S	S	P, D, S
0.8	S	S	S	D, S	S
0.9	S	S	D, S	D, S	S
1.0	S	S	S	S	S

3. Discussion

3.1. Collision Energetics

Figure 3 shows the time evolution of kinetic and potential energies of the two nanotubes bouncing back (B), connecting heterojunction (C) and fusing into a perfect BCN nanotube (P) after collision. It can be seen that all kinetic energy curves first drop sharply, then fluctuate up and down, and finally reach equilibrium. Because the total energy of the system remains constant, the evolution behavior of the corresponding potential energy is exactly opposite to that of the kinetic energy. In addition, for case B, the kinetic energy of the system after stabilization is lower than the initial kinetic energy, and the reduced kinetic energy has been transferred into the potential energy of the system, resulting in the increase in the potential energy, that is, the cylindrical tubular structure becomes an ellipsoidal cylinder after collision, as shown in Figure 3b and Video S1. For case C, the final kinetic energy is also lower than the initial kinetic energy. On the one hand, the reduced kinetic energy has been converted into the binding energy required for the formation of the junction between the two nanotubes, and also into the potential energy required for the deformation of the nanotubes, as shown in Figure 1a and Video S2. For case P, the final kinetic energy is slightly higher than the initial kinetic energy, while the final potential energy is slightly lower than the initial potential energy, indicating that the system is more stable after collision. Furthermore, it can be observed that the highest potential energy (see the marker a in Figure 3b) of case P is higher than those (markers b and c) of cases B and C, which means that the system for case P could have enough collision energy to overcome the fusion barrier of forming a BCN nanotube with a larger diameter, while it is difficult for cases B and C to fuse to form a nanotube.

3.2. Electronic Structures

Figure 4 shows the band structures and density of states (DOS) of the (6,0) single-walled CNT (SWCNT), (6,0) single-walled BNNT (SWBNNT), (12,0) single-walled BCN nanotube (SWBCNNT), graphene nanoribbon (GNR), h-BN nanoribbon (BNNR) and BCN heteronanoribbon. It can be seen that the energy levels of (6,0) SWCNT cross the Fermi level (see Figure 4a), which agrees with the consensus that the (6,0) SWCNT is a semimetal [42]. The band gap of (6,0) SWBNNT is calculated to be 2.791 eV (see Figure 4b), in good agreement with the results calculated by Rubio et al. [43,44]. The valence band maximum (VBM) of the (6,0) single-walled BN nanotube is mainly contributed to by N atoms, while the conduction band minimum (CBM) is mainly composed of B atoms (See Figure 4b). The (12,0) single-walled BCN nanotube is a direct band-gap semiconductor with a band gap of 0.808 eV. The VBM is mainly contributed to by C and N atoms, and the CBM is mainly contributed to by C and B atoms (see Figure 4c). The band gap of GNR is close to 0 eV, which is consistent with the results reported by Zhou et al. [45,46]. The band gap of BNNR is calculated to be 4.536 eV, as shown in Figure 4e, which is in good agreement with the results reported by Leite et al. [47]. The VBM of BNNR is mainly composed of N atoms, while the CBM is mainly composed of B atoms. The BCN heteronanoribbon is a direct band-gap semiconductor with a band gap of 0.544 eV (see Figure 4f). The VBM of the BCN heteronanoribbon is composed of C and N atoms, and the CBM is mainly contributed to by C atoms, indicating that C atoms are the main factor that constitutes the small band gap of the BCN heteronanoribbon.

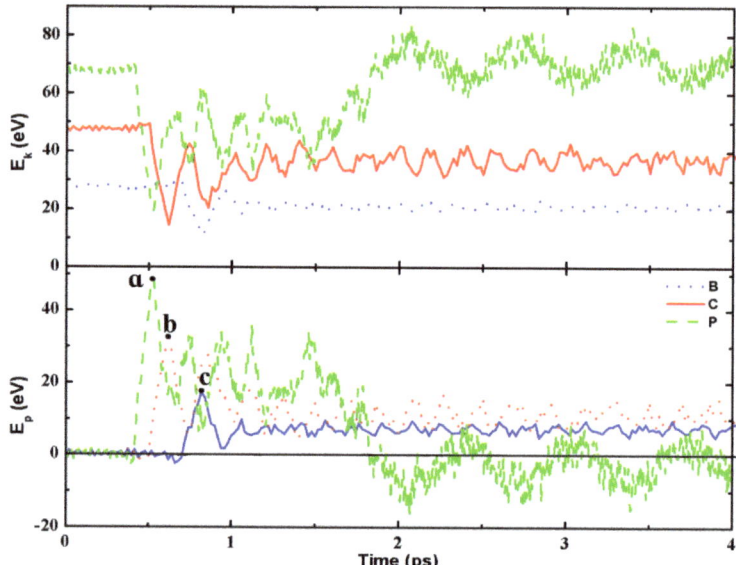

Figure 3. (Color online.) Time evolutions of the kinetic and potential energies of the two nanotubes bouncing back (B), connecting heterojunction (C) and fusing into a perfect BCN nanotube (P) after collision. The solid black line represents the total energy of the system, namely the sum of the kinetic and potential energies.

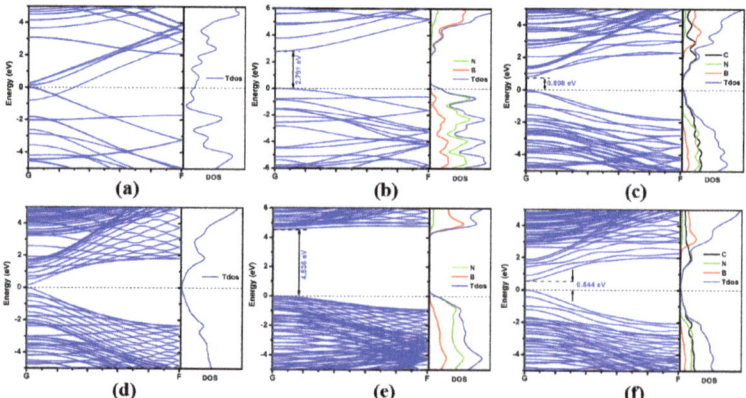

Figure 4. Energy bands and density of states of (6,0) single-walled CNT (**a**), (6,0) single-walled BNNT (**b**), (12,0) single-walled BCN heteronanotube (**c**), GNR (**d**), BNNR (**e**) and BCN heteronanoribbon (**f**).

The above calculation results indicate that the electronic properties of CNT/BNNT can be regulated effectively by the collision fusion of CNT and BNNT. In particular, the heteronanotubes and heteronanoribbons formed are new types of semiconductor nanomaterials, which may be applied in electronic devices, photocatalysis, integrated circuits and field effect transistors.

3.3. Defects

In these fusion structures attained, some of them contain defects, including vacancy, void, 5–7 pair defect, SW defect, ISW defect and 4–8 ring defect, which can have significant effects on the electrical, chemical and mechanical properties of the fusion structures. It was found that the formation of the 5–7 pair defect is conducive to the fusion of carbon nanotubes [48,49]. Lambin et al. [48] found that the 5–7 pair defects enable the two carbon nanotubes with different helicities to join and form molecular junctions, which had electrical properties different from the original nanotubes. Terrones et al. [49] found that if both carbon nanotubes with the same helicity contain a 5–7 pair defect, a CNT with a larger diameter can be formed through the fusion of the two nanotubes.

Among these defects, the 4–8 ring defect is the most easily formed defect structure after a CNT colliding with a BNNT. During the formation processes of BCN heteronanotube with 4–8 ring defects, four-membered rings containing C-C, C-N, C-B and B-N bonds are formed first, but then they all evolve into four-membered rings containing B-N bonds, as shown in Figure 5a,b. Based on our first-principles calculations, it is found that the energy of the BCN nanotube with B-N-B-N structure is lower than that with C-C-B-N structure. Therefore, it is easy to form four-membered ring structures with all B-N bonds in the evolution process. Figure 5c shows the electronic structure of the BCN nanotube with 4–8 ring defects. As can be seen, the heterostructure is a direct band-gap semiconductor with a band gap of 0.565 eV, which is slightly lower than that of the defect-free BCN heteronanotube. This indicates that the formation of the defects could effectively regulate the properties of the heteronanotube.

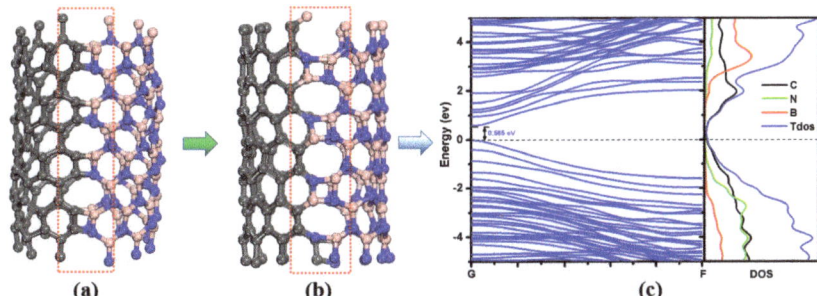

Figure 5. Schematic diagram of BCN heteronanotubes with alternating connections of four- and eight-membered rings. (**a**) Four-membered rings with C-C, C-B, B-N and N-C bonds. (**b**) Four-membered rings with each bond being a B-N type. (**c**) Energy band and density of states corresponding to the structure in (**b**).

In addition, some experimental techniques may be used to reduce the defects in the fused nanostructures. For example, defects can be eliminated by a combination of quenching and annealing methods. Previous SCC-DFTB simulations on the quenching and annealing processes of CNTs and GNRs with defects under a high temperature environment [50] indeed showed that the defects in both CNTs and GNRs could be substantially reduced. Therefore, a combination of quenching and annealing methods can be an effective approach to eliminate defects or reduce the defect density in heterostructures obtained by collision fusion.

4. Computational Methods

Collision dynamics simulations were implemented with the DFTB+ open source package [51–53]. Both CNTs and BNNTs were single-walled nanotubes with a chirality index of (6,0) and a length of about 17.62 Å, containing 192 atoms in total. The initial collision energy (E_k) ranged from 0.1 eV/atom to 1.0 eV/atom with an energy interval of 0.1 eV/atom (the collision energy was calculated from carbon atoms to ensure that the initial collision velocities of the two tubes are the same, and the corresponding collision velocity was varied from 12.67 Å/ps to 40.07 Å/ps). The atomic motion followed through Newton's equation, which was solved using the velocity Verlet algorithm [54]. The timestep was set to be 1 fs to ensure the conservations of momentum and energy. The total evolution time was 10 ps. The same collision results could be obtained after testing a longer simulation time. The initial temperature of the system was set at 300 K to investigate the collision fusion of CNTs and BNNTs at room temperature. In addition, it is noted that the initial structures of both nanotubes have been first optimized and then relaxed at 300 K, which corresponds to the initial setup of atomic velocities. The initial atomic velocities are very small and have no effect on the collision velocity. Molecular dynamics simulations were implemented in the microcanonical NVE ensemble [55].

All calculations in this paper focused on the parallel collisions of nanotubes, as shown in Figure 6. Considering the symmetry of nanotubes, five collision modes were considered to investigate the effects of collision locations on the collision results. As shown in Figure 7, a string of atoms along the axis in the CNT collided rightly with a string of atoms in the BNNT, the atoms and the centers of hexagonal rings, as well as the centers of bonds, i.e., atoms-colliding-atoms, atoms-colliding-atoms-hexagons, atoms-colliding-bonds, which were denoted as AA, AAH and AB, respectively. On the basis of type AA, CNT remained unchanged, and BN nanotubes were rotated by 15° and 30°, which were called h-AH and AH, respectively. In addition, five collision simulations were carried out for each of the five collision modes, and the collision results were statistically averaged for all collision simulations.

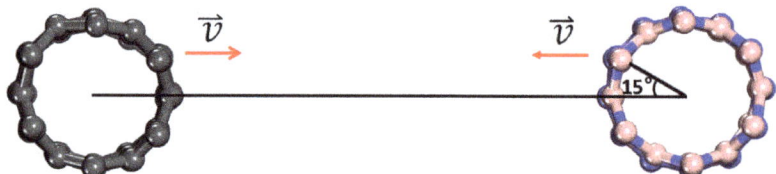

Figure 6. (Color online.) Schematic diagram of the collision setup for (6,0) single-walled CNT (Left) and (6,0) single-walled BNNT (Right) with the same initial velocity but in opposite directions. Gray, blue and pink balls represent carbon, nitrogen and boron atoms, respectively.

In order to ensure that there was no interaction between the two nanotubes at the beginning of the simulations, the initial distance between the walls of two nanotubes was set to be at least 20.00 Å. A periodic boundary condition along the tube axis was applied to simulate a nanotube with infinite length. Considering the presence of boron, carbon and nitrogen atoms in the system, the self-consistent-charge method was performed to avoid undesired electronic states of zwitterions [56].

The first-principles software package CASTEP was used to optimize the fused structure and calculate the electronic structure [57]. The generalized gradient approximation (GGA) parameterized by Perdew, Burke and Ernzerhof (PBE) was applied for the exchange and correlation interactions [58]. The cut-off energy for the plane-wave expansion was set to be 500 eV and the k-point was chosen as 0.01 Å^{-1}. The force on each ion was less than 0.01 eV/Å and the energy was converged within 1.0×10^{-5} eV/atom.

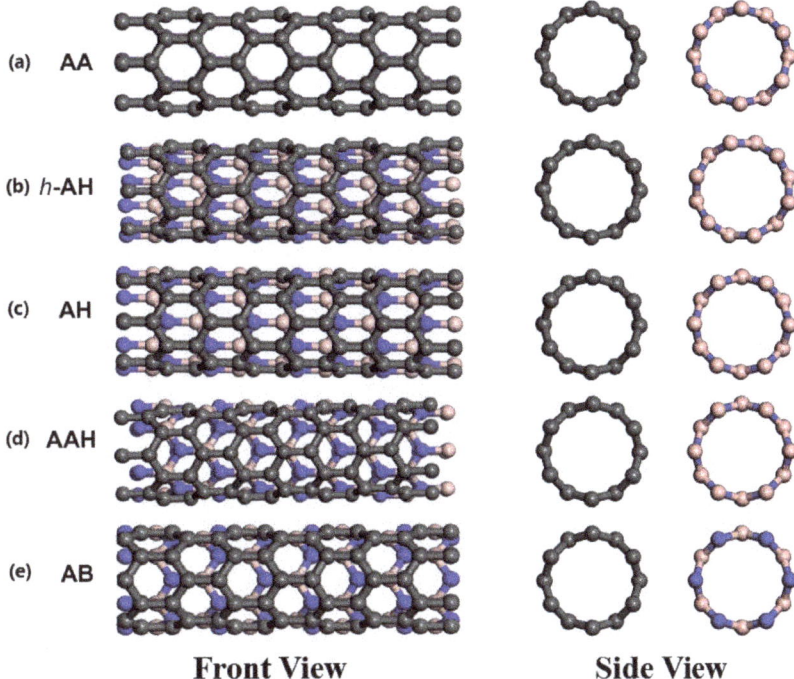

Figure 7. (Color online.) Five collision modes. (**a**) AA, (**b**) *h*-AH, (**c**) AH, (**d**) AAH and (**e**) AB.

5. Conclusions

In this work, the collision dynamics between CNTs and BNNTs under five parallel collision modes of AA, *h*-AH, AH, AAH and AB were investigated by the self-consistent-charge density-functional tight-binding molecular dynamics method. The five main collision outcomes are: the two nanotubes can (1) bounce back, (2) collide into heterojunctions, (3) fuse into BCN heteronanotubes, (4) form BCN heteronanoribbons and (5) collide to cause serious damage after collision. There may be defects created in the fused structure, including vacancy, void and irregular defects such as the 5–7 pair defect, SW defect, ISW defect and 4–8 ring defects. With the increase in collision energy, the number of defects increases, and the nanotubes may even break up into single atoms, dimers, trimers or chains of atoms. In addition, the electronic structures of BCN heteronanotubes and BCN heteronanoribbons formed by collision were investigated based on first-principles calculations. It was found that both the defect-free (12,0) BCN heteronanotube and the BCN heteronanoribbon are direct band-gap semiconductors with band gaps of 0.808 eV and 1.34 eV, respectively. These results indicate that the electronic structures of nanotubes can be effectively tuned by collision of CNTs and BNNTs, which could have significant implications in the field of electronic devices, photocatalysis, integrated circuits and field effect transistors. Although the present work only focuses on the collision fusion of CNTs and BNNTs with specific helicity, it can be extended to the collision processes of nanotubes with any helicity. Therefore, the present work not only theoretically demonstrates a novel method to create a heterostructure via collision fusion but also gains in-depth understanding in the synthesis of heteronanotubes and heteronanoribbons that can guide experiments.

Supplementary Materials: The following supporting information can be downloaded at https://www.mdpi.com/article/10.3390/molecules28114334/s1, Video S1: Bouncing off of two nanotubes after collision. Video S2: Merging together of two nanotubes after collision. Video S3: Forming BCN nanotube with no defect. Video S4: Forming BCN nanotube with defects.

Author Contributions: Conceptualization, C.Z. and Y.-W.Z.; methodology, C.Z.; software, J.X.; validation, K.R., H.S. and Z.G.Y.; formal analysis, C.Z.; investigation, K.R. and J.X.; resources, Y.-W.Z.; data curation, C.Z.; writing—original draft preparation, C.Z.; writing—review and editing, K.R.; visualization, J.X.; supervision, Y.-W.Z.; project administration, H.S.; funding acquisition, C.Z. and Y.-W.Z. All authors have read and agreed to the published version of the manuscript.

Funding: This research was funded by the Anhui Provincial Natural Science Foundation under Grant No. 2108085MA25, the Singapore national Research Foundation under Award No. NRF-CRP24-2020-0002, and Kai Ren thanks the support of the Natural Science Foundation of Jiangsu (No. BK20220407).

Institutional Review Board Statement: Not applicable.

Informed Consent Statement: Not applicable.

Data Availability Statement: The data that support the findings of this study are available from the corresponding author upon reasonable request.

Acknowledgments: The authors thank the use of computing resources at the A*STAR Computational Resource Centre and National Supercomputer Centre, Singapore.

Conflicts of Interest: The authors declare that they have no known competing financial interests or personal relationships that could have appeared to influence the work reported in this paper.

References

1. Wu, H.; Wang, Y.J.; Xu, Y.F.; Sivakumar, P.K.; Pasco, K.; Filippozzi, U.; Parkin, S.S.P.; Zeng, Y.J.; McQueen, T.; Ali, M.N. The field-free Josephson diode in a van der Waals heterostructure. *Nature* **2022**, *604*, 653–656. [CrossRef] [PubMed]
2. Ren, W.J.; Ouyang, Y.L.; Jiang, P.F.; Yu, C.Q.; He, J.; Chen, J. The Impact of Interlayer Rotation on Thermal Transport Across Graphene/Hexagonal Boron Nitride van der Waals Heterostructure. *Nano Lett.* **2021**, *21*, 2634–2641. [CrossRef]
3. Li, Y.; Zhang, J.W.; Chen, Q.G.; Xia, X.H.; Chen, M.H. Emerging of Heterostructure Materials in Energy Storage: A Review. *Adv. Mater.* **2021**, *33*, 2100855. [CrossRef] [PubMed]
4. Huang, S.Z.; Wang, Z.H.; Lim, Y.V.; Wang, Y.; Li, Y.; Zhang, D.H.; Yang, H.Y. Recent Advances in Heterostructure Engineering for Lithium–Sulfur Batteries. *Adv. Energy Mater.* **2021**, *11*, 2003689. [CrossRef]
5. Tang, F.H.; He, D.X.; Jiang, H.; Wang, R.S.; Li, Z.L.; Xue, W.D.; Zhao, R. The coplanar graphene oxide/graphite heterostructure-based electrodes for electrochemical supercapacitors. *Carbon* **2022**, *197*, 163–170. [CrossRef]
6. Yuan, K.; Hao, P.J.; Zhou, Y.; Hu, X.C.; Zhang, J.B.; Zhong, S.W. A two-dimensional MXene/BN van der Waals heterostructure as an anode material for lithium-ion batteries. *Phys. Chem. Chem. Phys.* **2022**, *24*, 13713–13719. [CrossRef]
7. Liu, C.; Lu, Y.H.; Yu, X.T.; Shen, R.J.; Wu, Z.M.; Yang, Z.S.; Yan, Y.F.; Feng, L.X.; Lin, S.S. Hot carriers assisted mixed-dimensional graphene/MoS$_2$/p-GaN light emitting diode. *Carbon* **2022**, *197*, 192–199. [CrossRef]
8. Wang, B.A.; Yuan, H.K.; Yang, T.; Wang, P.; Xu, X.H.; Chang, J.L.; Kuang, M.Q.; Chen, H. A two-dimensional PtS2/BN heterostructure as an S-scheme photocatalyst with enhanced activity for overall water splitting. *Phys. Chem. Chem. Phys.* **2022**, *24*, 26908–26914. [CrossRef]
9. Song, J.; Jiang, M.J.; Wan, C.; Li, H.J.; Zhang, Q.; Chen, Y.H.; Wu, X.H.; Yin, X.M.; Liu, J.F. Defective graphene/SiGe heterostructures as anodes of Li-ion batteries: A first-principles calculation study. *Phys. Chem. Chem. Phys.* **2022**, *25*, 617–624. [CrossRef]
10. Sun, X.X.; Zhu, C.G.; Yi, J.L.; Xiang, L.; Ma, C.; Liu, H.W.; Zheng, B.Y.; Liu, Y.; You, W.X.; Zhang, W.J.; et al. Reconfigurable logic-in-memory architectures based on a two-dimensional van der Waals heterostructure device. *Nat. Electron.* **2022**, *5*, 752–760. [CrossRef]
11. Zheng, Q.; Zhuang, Y.C.; Sun, Q.F.; He, L. Coexistence of electron whispering-gallery modes and atomic collapse states in graphene/WSe2 heterostructure quantum dots. *Nat. Commun.* **2022**, *13*, 1597. [CrossRef] [PubMed]
12. Chepkasov, I.V.; Smet, J.H.; Krasheninnikov, A.V. Single- and Multilayers of Alkali Metal Atoms inside Graphene/MoS$_2$ Heterostructures: A Systematic First-Principles Study. *J. Phys. Chem. C* **2022**, *126*, 15558–15564. [CrossRef]
13. Yang, Z.H.; Wu, M.S.; Luo, W.W.; Liu, G.; Xu, B. Structural, Electronic, and Transport Properties of Phosphorene–Graphene Lateral Heterostructure Anodes: Insights from First-Principles Calculations. *J. Phys. Chem. C* **2022**, *126*, 8928–8937. [CrossRef]
14. Kuang, H.F.; Zhang, H.Q.; Liu, X.H.; Chen, Y.D.; Zhang, W.G.; Chen, H.; Ling, Q.D. Microwave-assisted synthesis of NiCo-LDH/graphene nanoscrolls composite for supercapacitor. *Carbon* **2022**, *190*, 57–67. [CrossRef]
15. Wang, H.; Chen, J.M.; Lin, Y.P.; Wang, X.H.; Li, J.M.; Li, Y.; Gao, L.J.; Zhang, L.B.; Chao, D.L.; Xiao, X.; et al. Electronic Modulation of Non-van der Waals 2D Electrocatalysts for Efficient Energy Conversion. *Adv. Mater.* **2021**, *33*, 2008422. [CrossRef] [PubMed]

16. Song, S.; Gong, J.; Jiang, X.W.; Yang, S.Y. Influence of the interface structure and strain on the rectification performance of lateral MoS$_2$/graphene heterostructure devices. *Phys. Chem. Chem. Phys.* **2022**, *24*, 2265–2274. [CrossRef]
17. Iijima, S. Helical microtubules of graphitic carbon. *Nature* **1991**, *354*, 56–58. [CrossRef]
18. Han, L.Y.; Xiao, C.X.; Song, Q.; Yin, X.M.; Li, W.; Li, K.Z.; Li, Y.Y. Nano-interface effect of graphene on carbon nanotube reinforced carbon/carbon composites. *Carbon* **2022**, *190*, 422–429. [CrossRef]
19. Zhang, S.; Pang, J.B.; Li, Y.F.; Yang, F.; Gemming, T.; Wang, K.; Wang, X.; Peng, S.G.; Liu, X.Y.; Chang, B.; et al. Emerging Internet of Things driven carbon nanotubes-based devices. *Nano Res.* **2022**, *15*, 4613–4637. [CrossRef]
20. Wang, R.R.; Wu, R.B.; Yan, X.X.; Liu, D.; Guo, P.F.; Li, W.; Pan, H.G. Implanting Single Zn Atoms Coupled with Metallic Co Nanoparticles into Porous Carbon Nanosheets Grafted with Carbon Nanotubes for High-Performance Lithium-Sulfur Batteries. *Adv. Funct. Mater.* **2022**, *32*, 2200424. [CrossRef]
21. Zhu, R.F.; Wang, D.; Liu, Y.M.; Liu, M.M.; Fu, S.H. Bifunctional superwetting carbon nanotubes/cellulose composite membrane for solar desalination and oily seawater purification. *Chem. Eng. J.* **2022**, *433*, 133510. [CrossRef]
22. Poggioli, A.R.; Limmer, D.T. Distinct Chemistries Explain Decoupling of Slip and Wettability in Atomically Smooth Aqueous Interfaces. *J. Phys. Chem. Lett.* **2021**, *12*, 9060–9067. [CrossRef] [PubMed]
23. Stern, H.L.; Gu, Q.S.; Jarman, J.; Barker, S.E.; Mendelson, N.; Chugh, D.; Schott, S.; Tan, H.H.; Sirringhaus, H.; Aharonovich, I.; et al. Room-temperature optically detected magnetic resonance of single defects in hexagonal boron nitride. *Nat. Commun.* **2022**, *13*, 618. [CrossRef] [PubMed]
24. Pham, P.V.; Bodepudi, S.C.; Shehzad, K.; Liu, Y.; Xu, Y.; Yu, B.; Duan, X.F. 2D Heterostructures for Ubiquitous Electronics and Optoelectronics: Principles, Opportunities, and Challenges. *Chem. Rev.* **2022**, *122*, 6514–6613. [CrossRef]
25. Turiansky, M.E.; Alkauskas, A.; van de Walle, C.G. Spinning up quantum defects in 2D materials. *Nat. Mater.* **2020**, *19*, 487–489. [CrossRef]
26. Mirzayev, M.N. Heat transfer of hexagonal boron nitride (h-BN) compound up to 1 MeV neutron energy: Kinetics of the release of wigner energy. *Radiat. Phys. Chem.* **2021**, *180*, 109244. [CrossRef]
27. Xu, T.; Zhang, K.; Cai, Q.R.; Wang, N.Y.; Wu, L.Y.; He, Q.; Wang, H.; Zhang, Y.; Xie, Y.F.; Yao, Y.G.; et al. Advances in synthesis and applications of boron nitride nanotubes: A review. *Chem. Eng. J.* **2022**, *431*, 134118. [CrossRef]
28. Qi, R.S.; Li, N.; Du, J.L.; Shi, R.C.; Huang, Y.; Yang, X.X.; Liu, L.; Xu, Z.; Dai, Q.; Yu, D.P.; et al. Four-dimensional vibrational spectroscopy for nanoscale mapping of phonon dispersion in BN nanotubes. *Nat. Commun.* **2021**, *12*, 1179. [CrossRef]
29. Konabe, S. Exciton effect on shift current in single-walled boron-nitride nanotubes. *Phys. Rev. B* **2021**, *103*, 075402. [CrossRef]
30. Feng, Y.; Li, H.N.; Hou, B.; Kataura, H.; Inoue, T.; Chiashi, S.; Xiang, R.; Maruyama, S. Zeolite-supported synthesis, solution dispersion, and optical characterizations of single-walled carbon nanotubes wrapped by boron nitride nanotubes. *J. Appl. Phys.* **2021**, *129*, 015101. [CrossRef]
31. Jones, R.S.; Maciejewska, B.; Grobert, N. Synthesis, characterisation and applications of core–shell carbon–hexagonal boron nitride nanotubes. *Nanoscale Adv.* **2020**, *2*, 4996–5014. [CrossRef]
32. Stephan, O.; Ajayan, P.M.; Colliex, C.; Redlich, P.; Lambert, J.M.; Bernier, P.; Lefin, P. Doping Graphitic and Carbon Nanotube Structures with Boron and Nitrogen. *Science* **1994**, *266*, 1683–1685. [CrossRef]
33. Wang, Y.; Huang, G.; Zhang, J.; Shao, Q.Y. Tunable electronic properties of ultra-thin boron-carbon-nitrogen heteronanotubes for various compositions. *J. Mol. Model.* **2014**, *20*, 2371–2378. [CrossRef] [PubMed]
34. Chaudhuri, P.; Lima, C.N.; Frota, H.O.; Ghosh, A. First-principles study of nanotubes of carbon, boron and nitrogen. *Appl. Surf. Sci.* **2019**, *490*, 242–250. [CrossRef]
35. Belgacem, A.B.; Hinkov, I.; Yahia, S.B.; Brinza, O.; Farhat, S. Arc discharge boron nitrogen doping of carbon nanotubes. *Mater. Today Commun.* **2016**, *8*, 183–195. [CrossRef]
36. Campbell, E.E.B.; Schyja, V.; Ehlich, R.; Hertel, I.V. Observation of molecular fusion and deep inelastic scattering in C_{60}^{2+} + C_{60} collisions. *Phys. Rev. Lett.* **1993**, *70*, 263. [CrossRef] [PubMed]
37. Rohmund, F.; Campbell, E.E.B. Charge transfer collisions between C_{60}^{2+} and C_{60}. *Chem. Phys. Lett.* **1995**, *245*, 237–243. [CrossRef]
38. Rohmund, F.; Glotov, A.; Hansen, K.; Campbell, E.E.B. Experimental studies of fusion and fragmentation of fullerenes. *J. Phys. B* **1996**, *29*, 5143–5161. [CrossRef]
39. Lee, G.D.; Wang, C.Z.; Yoon, E.; Hwang, N.M.; Ho, K.M. The role of pentagon–heptagon pair defect in carbon nanotube: The center of vacancy reconstruction. *Appl. Phys. Lett.* **2010**, *97*, 093106. [CrossRef]
40. Stone, A.J.; Wales, D.J. Theoretical studies of icosahedral C_{60} and some related species. *Chem. Phys. Lett.* **1986**, *128*, 501–503. [CrossRef]
41. Lusk, M.T.; Wu, D.T.; Carr, L.D. Graphene nanoengineering and the inverse Stone-Thrower-Wales defect. *Phys. Rev. B* **2010**, *81*, 155444. [CrossRef]
42. Odom, T.W.; Huang, J.L.; Kim, P.; Lieb, C.M. Atomic structure and electronic properties of single-walled carbon nanotubes. *Nature* **1998**, *391*, 62–64. [CrossRef]
43. Rubio, A.; Corkill, J.L.; Cohen, M.L. Theory of graphitic boron nitride nanotubes. *Phys. Rev. B* **1994**, *49*, 5081–5084. [CrossRef] [PubMed]
44. Blasé, X.; Rubio, A.; Louie, S.G.; Cohen, M.L. Stability and Band Gap Constancy of Boron Nitride Nanotubes. *Europhys. Lett.* **1994**, *28*, 335–340. [CrossRef]

45. Zhou, S.Y.; Gweon, G.H.; Fedorov, A.V.; First, P.N.; De Heer, W.A.; Lee, D.H.; Guinea, F.; Castro Neto, A.H.; Lanzara, A. Substrate-induced bandgap opening in epitaxial graphene. *Nat. Mater.* **2007**, *6*, 770–775. [CrossRef]
46. Son, Y.W.; Cohen, M.L.; Louie, S.G. Energy Gaps in Graphene Nanoribbons. *Phys. Rev. Lett.* **2006**, *97*, 216803. [CrossRef] [PubMed]
47. Zhang, Z.H.; Guo, W.L. Energy-gap modulation of BN ribbons by transverse electric fields: First-principles calculations. *Phys. Rev. B* **2008**, *77*, 075403. [CrossRef]
48. Lambin, P.; Vigneron, J.P.; Fonseca, A.; Nagy, J.B.; Lucas, A.A. Atomic structure and electronic properties of a bent carbon nanotube. *Synth. Met.* **1996**, *77*, 249–252. [CrossRef]
49. Terrones, M.; Terrones, H.; Banhart, F.; Charlier, J.-C.; Ajayan, P.M. Coalescence of Single-Walled Carbon Nanotubes. *Science* **2000**, *288*, 1226–1229. [CrossRef]
50. Zhang, C.; Mao, F.; Meng, X.R.; Wang, D.Q.; Zhang, F.S. Collision-induced fusion of two single-walled carbon nanotubes: A quantitative study. *Chem. Phys. Lett.* **2016**, *657*, 184–189. [CrossRef]
51. Aradi, B.; Hourahine, B.; Frauenheim, T. DFTB+, a Sparse Matrix-Based Implementation of the DFTB Method. *J. Phys. Chem. A* **2007**, *111*, 5678–5684. [CrossRef] [PubMed]
52. Enyashin, A.N.; Ivanovskii, A.L. Mechanical and electronic properties of a C/BN nanocable under tensile deformation. *Nanotechnology* **2005**, *16*, 1304–1310. [CrossRef]
53. Enyashin, A.N.; Seifert, G.; Ivanovskii, A.L. Calculation of the Electronic and Thermal Properties of C/BN Nanotubular Heterostructures. *Inorg. Mater.* **2005**, *41*, 595–603. [CrossRef]
54. Swope, W.C.; Andersen, H.C.; Berens, P.H.; Wilson, K.R. A computer simulation method for the calculation of equilibrium constants for the formation of physical clusters of molecules: Application to small water clusters. *J. Chem. Phys.* **1982**, *76*, 637–649. [CrossRef]
55. Kraska, T. Molecular-dynamics simulation of argon nucleation from supersaturated vapor in the NVE ensemble. *J. Chem. Phys.* **2006**, *124*, 054507. [CrossRef] [PubMed]
56. Jakowski, J.; Irle, S.; Morokuma, K. Collision-induced fusion of two C_{60} fullerenes: Quantum chemical molecular dynamics simulations. *Phys. Rev. B* **2010**, *82*, 125443. [CrossRef]
57. Segall, M.D.; Lindan, P.J.D.; Probert, M.J.; Pickard, C.J.; Hasnip, P.J.; Clark, S.J.; Payne, M.C. First-principles simulation: Ideas, illustrations and the CASTEP code. *J. Phys. Condens. Matter* **2002**, *14*, 2717–2744. [CrossRef]
58. Perdew, J.P.; Burke, K.; Ernzerhof, M. Generalized Gradient Approximation Made Simple. *Phys. Rev. Lett.* **1996**, *77*, 3865–3868. [CrossRef]

Disclaimer/Publisher's Note: The statements, opinions and data contained in all publications are solely those of the individual author(s) and contributor(s) and not of MDPI and/or the editor(s). MDPI and/or the editor(s) disclaim responsibility for any injury to people or property resulting from any ideas, methods, instructions or products referred to in the content.

Communication

Ultrahigh Carrier Mobility in Two-Dimensional IV–VI Semiconductors for Photocatalytic Water Splitting

Zhaoming Huang [1,2], Kai Ren [2,3], Ruxin Zheng [4], Liangmo Wang [1,*] and Li Wang [3,5]

[1] School of Mechanical Engineering, Nanjing University of Science and Technology, Nanjing 210094, China
[2] School of Mechanical and Electronic Engineering, Nanjing Forestry University, Nanjing 211189, China; kairen@njfu.edu.cn (K.R.)
[3] School of Mechanical Engineering, Wanjiang University of Technology, Ma'anshan 243031, China
[4] School of Mechanical Engineering, Southeast University, Nanjing 211189, China
[5] Office of Academic Affairs, Xuancheng Vocational and Technical College, Xuancheng 242000, China
* Correspondence: liangmo@njust.edu.cn

Abstract: Two-dimensional materials have been developed as novel photovoltaic and photocatalytic devices because of their excellent properties. In this work, four δ-IV–VI monolayers, GeS, GeSe, SiS and SiSe, are investigated as semiconductors with desirable bandgaps using the first-principles method. These δ-IV–VI monolayers exhibit exceptional toughness; in particular, the yield strength of the GeSe monolayer has no obvious deterioration at 30% strain. Interestingly, the GeSe monolayer also possesses ultrahigh electron mobility along the x direction of approximately 32,507 $cm^2 \cdot V^{-1} \cdot s^{-1}$, which is much higher than that of the other δ-IV–VI monolayers. Moreover, the calculated capacity for hydrogen evolution reaction of these δ-IV–VI monolayers further implies their potential for applications in photovoltaic and nano-devices.

Keywords: IV–VI monolayers; mechanical property; carrier mobility; hydrogen evolution reaction

Citation: Huang, Z.; Ren, K.; Zheng, R.; Wang, L.; Wang, L. Ultrahigh Carrier Mobility in Two-Dimensional IV–VI Semiconductors for Photocatalytic Water Splitting. *Molecules* 2023, *28*, 4126. https://doi.org/10.3390/molecules28104126

Academic Editor: Lin Ju

Received: 28 March 2023
Revised: 10 May 2023
Accepted: 14 May 2023
Published: 16 May 2023

Copyright: © 2023 by the authors. Licensee MDPI, Basel, Switzerland. This article is an open access article distributed under the terms and conditions of the Creative Commons Attribution (CC BY) license (https://creativecommons.org/licenses/by/4.0/).

1. Introduction

Since graphene was proposed as a representative of two-dimensional (2D) materials in 2004 [1], considerable efforts have been made to explore its excellent mechanical [2], electronic [3], magnetic [4] and chemical [5] properties, showing promise for its widespread use in nano-devices. Since then, other 2D materials have also been studied by researchers [4,6–10]. Among them, transition metal dichalcogenides (TMDs) have also attracted great interest due to their novel properties [11,12], such as WSe_2, MoSSe and MoS_2. Monolayered WSe_2 possesses outstanding mechanical properties, and experiments have shown that when the layers of the WSe_2 is increased to 2–4, the photoluminescence (PL) is significantly enhanced by 2% under tensile strain [13]. It is reported that the carrier mobility of the MoS_2 monolayer is equivalent to that of its nanobelts. In addition, the carrier mobility in MoS_2 nanobelts can be made more robust while reducing the size of the monolayered MoS_2 [14]. The TMD materials with a Janus structure further result in an excellent performance [15]. MoSSe shows unusual properties in adsorption, with high gas sensitivity and surface and strain selectivity [16]. In order to expand the applications of 2D materials, the stacking of two different materials into a heterostructure using van der Waals (vdWs) force is a popular method. Based on this, different properties of materials can be induced at the interface. The Janus TMD vdW heterostructure also shows promise for photocatalytic and thermal applications [17–19].

The carrier mobility of 2D materials is a critical property used in the applications of nano-devices [20]. The carrier mobility of the CaP_3 monolayer is calculated to be 19,930 $cm^2 \cdot v^{-1} \cdot s^{-1}$, while the carrier mobility can be increased to be 22,380 $cm^2 \cdot v^{-1} \cdot s^{-1}$ by connecting two layers of CaP_3 together using vdWs forces [21]. A family of Li_xB_y monolayers were investigated using the evolutionary structure search method, and the monolayered Li_2B_6 showed a high hole mobility of approximately 6.8×10^3 $cm^2 \cdot v^{-1} \cdot s^{-1}$, showing that it can be used in high-speed electronic devices [22]. The high carrier mobility also contributes to the efficient photocatalytic properties in the hydrogen evolution reaction (HER). The ability of the HER is determined by the interaction between the intermediate and photocatalyst, which is evaluated using Gibbs free energy [23,24]. Importantly, the HER also can be tuned via defect [25,26], nanonization [27], heterostructures [28], etc. Moreover, the suitable band energy of the semiconductor is also critical in order to decompose the water so that the conduction band minimum (CBM) is more positive than −4.44 eV for the redox potentials (H^+/H_2) and the valence band maximum (VBM) is more negative than −5.67 eV for the oxidation potential (O_2/H_2O) at pH 0 [29]. More recently, the δ-IV–VI monolayers, GeS, GeSe, SiS and SiSe, were predicted to possess an excellent light absorption performance (even at up to 7.8×10^5 cm^{-1} for SiSe), meaning that they can be considered as promising photocatalysts. However, to be candidates for water splitting, the carrier mobility and the HER properties need to be further developed.

In this work, the density functional theory (DFT) is applied to systematically investigate the mechanical and electronic properties of the δ-IV–VI monolayers (GeS, GeSe, SiS and SiSe). Furthermore, the electronic and stress–strain responses are addressed. Next, the carrier mobility and the hydrogen evolution reaction of the δ-GeS, δ-GeS, δ-SiS and δ-SiSe monolayers are explored.

2. Results and Discussion

First, the structure of the δ-GeS, δ-GeS, δ-SiS and δ-SiSe monolayers were optimized as shown in Figure 1. The lattice parameters in the x (or y) direction were calculated as 5.58 Å (or 5.76 Å), 5.83 Å (or 5.81 Å), 5.50 Å (or 5.67 Å) and 5.69 Å (or 5.73 Å), respectively, for the δ-GeS, δ-GeS, δ-SiS and δ-SiSe monolayers. Moreover, the bond lengths between the Ge–S, Ge–S, Si–S and Si–Se monolayers were obtained as 2.42, 2.54, 2.32 and 2.44 Å, which are in good agreement with previous research [30]. These IV–VI monolayers possess a space group of $Pca2_1$, which was also reported in a previous investigation [30]. The cohesive energy (E_{co}) of the δ-IV–VI monolayers was calculated using $E_{co} = (4E_X + 4E_Y − E_{XY})/8$, where E_X, E_Y and E_{XY} are the total energies of a Ge (or Si) atom, a Se (or S) atom and the δ-IV–VI monolayer, respectively. The obtained cohesive energy of the GeS, GeSe, SiS and SiSe monolayers are 3.61, 3.37, 3.81 and 3.51 eV/atom, which are comparable with the values of phosphorene (approximately 3.48 eV/atom), germanene (approximately 3.24 eV/atom) and silicene (approximately 3.91 eV/atom) [31]. The obtained cohesive energy of SiS is larger than that of the puckered SiS (3.16 eV per atom) [32], and the obtained cohesive energy of the GeS and GeSe monolayers is also similar to that recently reported for GeS and GeSe with other phases [33–35]. Such IV–VI monolayers can be prepared in experiments using the chemical vapor deposition method and then isolated through mechanical, sonicated or liquid-phase exfoliation methods, which have been adopted to synthesize few-layer GaSe [36] and GeS [37].

The mechanical capacities of the δ-GeS, δ-GeS, δ-SiS and δ-SiSe monolayers were calculated by investigating the stress–strain response, as shown in Figure 2. The δ-GeS, δ-GeS, δ-SiS and δ-SiSe monolayers were more elastic in the x direction than the y direction. Through the linear fitting of the initial range (within 5%), the obtained Young's moduli (E) of the δ-GeS, δ-GeS, δ-SiS and δ-SiSe monolayers were defined as $E = \Delta Stress/\Delta Strain$, obtained to be approximately 34 $N \cdot m^{-1}$, 30 $N \cdot m^{-1}$, 39 $N \cdot m^{-1}$ and 28 $N \cdot m^{-1}$, respectively, along the x direction. Meanwhile, the Young's moduli are calculated as 21 $N \cdot m^{-1}$, 15 $N \cdot m^{-1}$, 17 $N \cdot m^{-1}$ and 20 $N \cdot m^{-1}$, respectively, along the y direction for the δ-GeS, δ-GeS, δ-SiS and δ-SiSe monolayers, in accordance with our other report [30]. The SiS and SiSe monolayers have yield strengths of approximately 3.13 $N \cdot m^{-1}$ and 2.72 $N \cdot m^{-1}$ at the strains of 15% and 18% in the x direction, respectively, as shown by the gray dashed lines in Figure 2c,d.

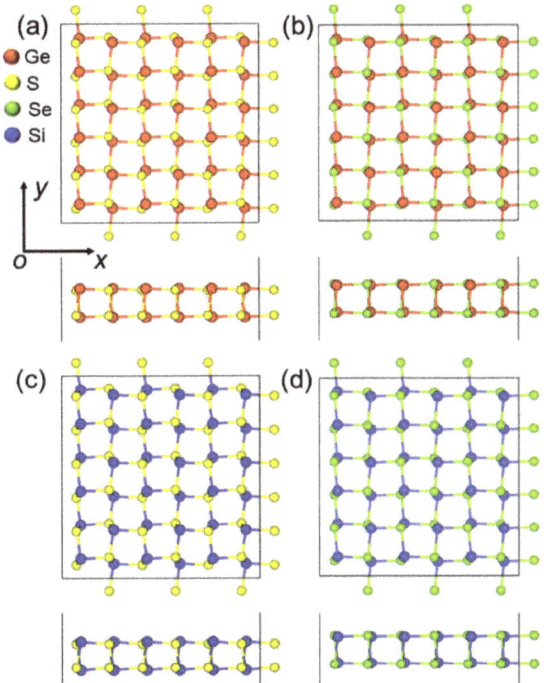

Figure 1. The top and side views of the atomic structures of the (**a**) GeS, (**b**) GeSe, (**c**) SiS and (**d**) SiSe monolayers.

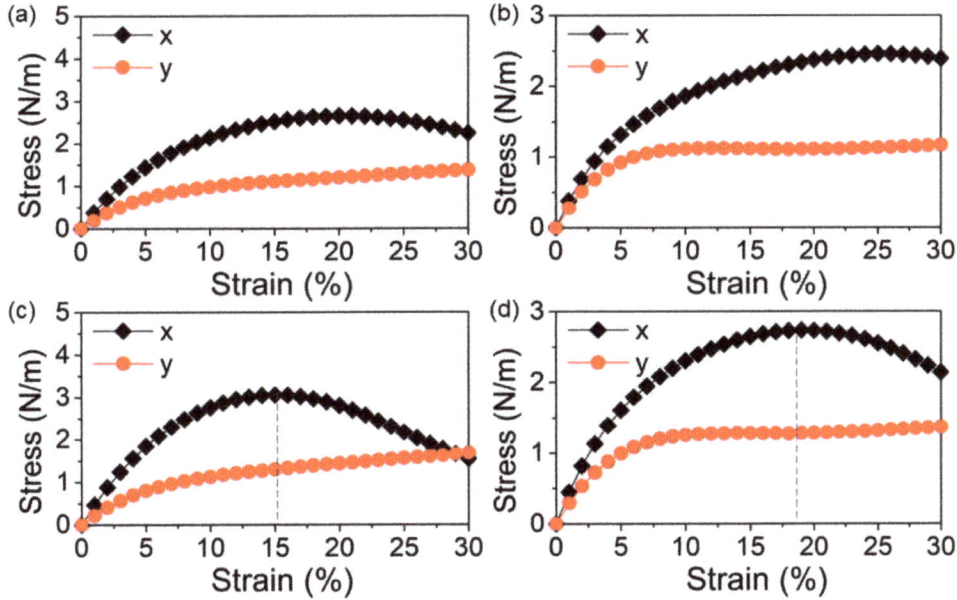

Figure 2. The strain–stress relationships of the (**a**) GeS, (**b**) GeSe, (**c**) SiS and (**d**) SiSe monolayers along the x and y directions.

The band structures of the δ-GeS, δ-GeS, δ-SiS and δ-SiSe monolayers shown in Figure 3 were determined using the PBE and HSE06 methods. One can see that all these monolayers are semiconductors, with bandgaps of approximately 2.65 eV (1.92 eV), 2.20 eV (1.60 eV), 2.15 eV (1.42 eV) and 2.08 eV (eV), being functional according to HSE06 (PBE). In Figure 3a–d, the δ-GeS, δ-GeS, δ-SiS and δ-SiSe monolayers possess almost exact bandgap structures, and the obtained bandgaps of these monolayers imply their decent application potential as photocatalysts for water splitting (larger than 1.23 eV) [38]. Furthermore, the band edge positions of these IV–VI monolayers are shown in Figure 3e at pH 0. Evidently, the δ-GeS, δ-GeS, δ-SiS and δ-SiSe monolayers demonstrate the sufficient energy of the CBM and VBM to induce the reductions and oxidations for water splitting. For comparative purposes, the band alignments of some TMD materials are also shown in Figure 3e, where one can see that only the WS_2 monolayer has a suitable band energy for redox.

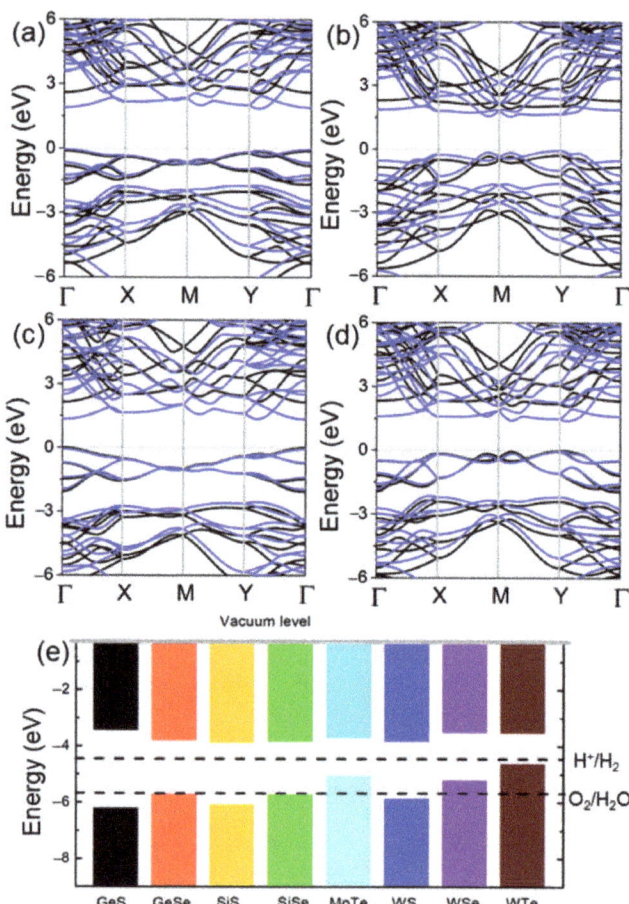

Figure 3. The DS-PAW calculated band structure of the (**a**) GeS, (**b**) GeSe, (**c**) SiS and (**d**) SiSe monolayers, (**e**) and the band alignment of these IV–VI monolayers compared with TMDs. The Fermi level is set as 0 eV. The blue and black lines represent the results of PBE and HSE06 calculations. The band edge energy was calculated with respect to the water oxidation (O_2/H_2O) and reduction (H^+/H_2) potentials at 0 pH.

Next, the carrier mobility of the δ-GeS, δ-GeS, δ-SiS and δ-SiSe monolayers is investigated, considering their decent bandgaps. The effective masses (m^*) of the electron and hole are determined by fitting the parabolic functions, which can be represented as:

$$m^* = \pm\hbar^2 \left(\frac{d^2 E_k}{dk^2}\right)^{-1}, \quad (1)$$

where k and E_k are the wave vector and the corresponding electronic energy, respectively. Furthermore, the carrier mobility (μ) of these 2D materials is calculated using:

$$\mu = \frac{e\hbar^3 C}{k_B T m^* m_e E_d^2}, \quad (2)$$

where the temperature is explained by T, e is the electron charge, the Planck constant is determined by $\hbar$, and k_B is the Boltzmann constant. The change in the band edge of these layered materials is evaluated using the deformation potential (E_d). It is worth noting that the obtained deformation potentials are compared based on the vacuum level. Moreover, the elastic modulus is used with C, which is obtained using $C = \left[\partial^2 E / \partial \varepsilon^2\right] / S$. Here, the total energy of the system is E and the area of the system is S. The energy differences among these δ-IV–VI monolayers are shown in Figure 4, and the fitted elastic moduli are summarized in Table 1.

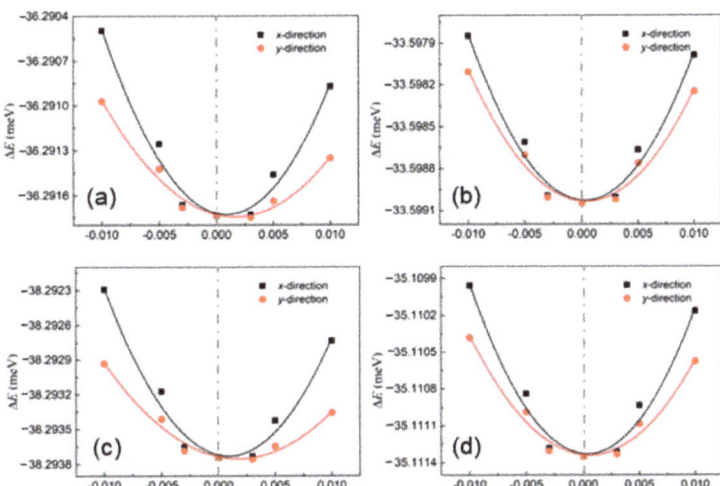

Figure 4. The energy differences among the (**a**) GeS, (**b**) GeSe, (**c**) SiS and (**d**) SiSe monolayers under different strains obtained by DS-PAW.

As an important parameter of carrier mobility, the deformation potential is calculated using the strain response to the band edge positions. For this purpose, the ranges of external and uniaxial strain are controlled within 0.01. The changes in the band edge positions under different strains on the δ-IV–VI monolayers are exhibited in Figure 5. One can see that the energy of the VBM for the GeSe and SiSe monolayers is more sensitive than that of the others, which implies obstructed hole mobility. The Bardeen–Shockley deformation potential theory is considered in the calculations for the strain effect on the energy and the band edge position, which can be used to explore long-range electrostatic terms in the theory of electronic deformation potential, with the results also showing good agreement with the experiments [39]. As the effective masses calculated using the HSE06 functional may be inaccurate due to the effect of Hartree−Fock exchange [40], the PBE functional is used to predict the carrier mobility.

Table 1. The obtained effective mass, elastic modulus, deformation potential constant and the carrier mobility of the hole (h) and the electron (e) for the GeS, GeSe, SiS and SiSe monolayers in the x and y directions using DFT calculations.

Materials	Direction	Carrier	m	E (eV)	C	μ
GeS	x	e^-	2.95	0.79	41	465
		h^+	−1.33	0.50		1246
	y	e^-	0.16	−1.61	23	1140
		h^+	−1.49	0.34		1312
GeSe	x	e^-	0.11	−0.92	43	32,507
		h^+	−1.32	5.10		45
	y	e^-	0.35	−0.85	33	9543
		h^+	−0.12	−4.82		439
SiS	x	e^-	1.24	−0.53	50	2041
		h^+	−1.04	−0.49		2489
	y	e^-	0.82	−1.39	25	220
		h^+	−1.33	0.74		411
SiSe	x	e^-	2.70	−1.04	51	319
		h^+	−1.28	−6.28		16
	y	e^-	0.22	−0.97	34	2997
		h^+	−0.66	−5.66		25

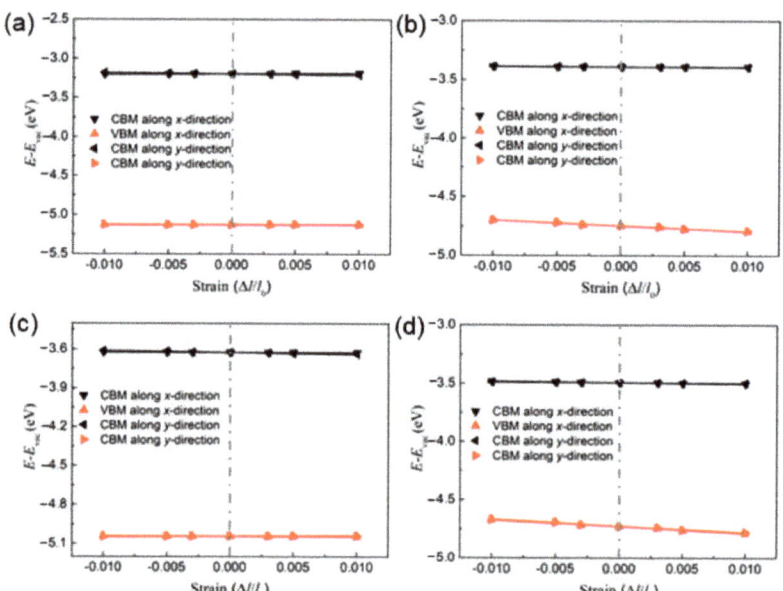

Figure 5. The changes in the band edge positions of the (**a**) GeS, (**b**) GeSe, (**c**) SiS and (**d**) SiSe monolayers under different strains obtained by DS-PAW.

The carrier mobility of these δ-IV–VI monolayers was calculated, as shown in Table 1. Interestingly, the carrier mobility of the SiS monolayer in the x direction is approximately 10 times higher than that in the y direction, showing favorable carrier transport direction along x [41]. Meanwhile, the carrier mobility of the electrons is much higher than that of the holes in the GeSe and SiSe monolayers, which is advantageous for the separation of the photogenerated electrons and holes [42]. More importantly, the GeSe monolayer possess an ultrahigh electron mobility in the x direction of approximately 32,507 cm$^2 \cdot$V$^{-1} \cdot$s^{-1}, which is higher than that of black phosphorus [43]. In addition, the other obtained carrier mobilities of the GeS (465–1312 cm$^2 \cdot$V$^{-1} \cdot$s^{-1}), SiS (2202–2489 cm$^2 \cdot$V$^{-1} \cdot$s^{-1}) and SiSe (2997 cm$^2 \cdot$V$^{-1} \cdot$s^{-1} for electron) are also higher than those of other novel 2D materials, such as WS$_2$ (542 cm$^2 \cdot$V$^{-1} \cdot$s^{-1}) [44], MoS$_2$ (201 cm$^2 \cdot$V$^{-1} \cdot$s^{-1}) [14], BSe (2396 cm$^2 \cdot$V$^{-1} \cdot$s^{-1}) [45], etc. Moreover, the obtained ultrahigh electron mobility of the GeSe monolayer in the x direction is attributed to its small deformation potential constant (about −0.92) and effective mass (approximately 0.11), suggesting the insensitivity of the band edge position to the external strain. Even though the GeS and SiS monolayers present small deformation potential constants, the carrier mobility is suppressed by the larger effective mass.

The catalytic properties of these IV−VI monolayers were also determined. The Gibbs free energy change (ΔG_{H^*}) of the GeS, GeSe, SiS and SiSe monolayers was investigated under standard conditions using:

$$\Delta G_{H^*} = \Delta E + \Delta E_{zpe} + T\Delta S, \tag{3}$$

where the total energy of the H-adsorbed IV−VI monolayers, the difference in the zero-point energies and the change in entropy caused by the adsorption are represented as ΔE, ΔE_{zpe} and ΔS, respectively. T is defined as 298.15 K. The active site is marked by the sign "*". The HER characteristic is induced via two reactions:

$$* + H^+ + e^- \rightarrow H^*, \tag{4}$$

$$H^* + H^+ + e^- \rightarrow H_2 + *. \tag{5}$$

Furthermore, the most favorable H-adsorbed sites of the systems are demonstrated in Figure 6a, and the calculated Gibbs free energies of these H-adsorbed GeS, GeSe, SiS and SiSe monolayers are obtained as −1.775 eV, 2.480 eV, 2.569 eV and 2.965 eV, respectively, as shown in Figure 6b. One can see that the GeS possesses an advantageous HER ability, which is even smaller than that of the MoSi$_2$N$_4$ (2.79 eV) and MoSi$_2$N$_4$ (2.51 eV) monolayers [46].

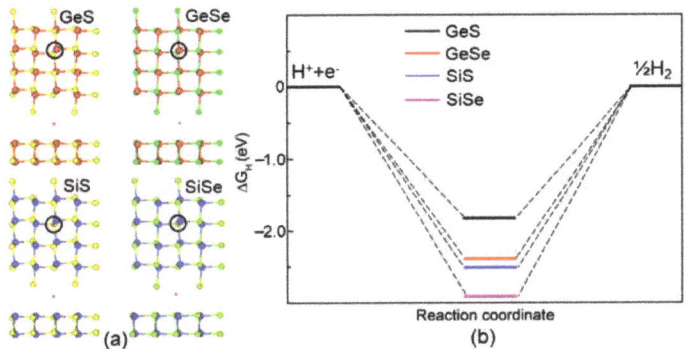

Figure 6. (a) The most favorable calculated adsorption sites and (b) the Gibbs free energies of the IV−VI monolayers obtained by DS-PAW.

3. Computational Methods

In our simulations, the calculations for the structural optimization, electronic property, carrier mobility and the HER performances were calculated by Device Studio [Hongzhiwei Technology, Device Studio, Version 2021A, China, 2021. Available online: https://iresearch.net.cn/cloudSoftware] program, which provides a number of functions for performing visualization, modeling and simulation. And all that simulations using DS-PAW software are integrated in Device Studio program [47]. All the mechanical calculations were conducted based on DFT using the first-principles method with the Vienna ab initio simulation package (VASP) [48]. Generalized gradient approximation (GGA) was employed together with projector augmented wave potentials (PAW) to demonstrate core electrons [49,50]. The Perdew–Burke–Ernzerhof (PBE) functional was used to explain the exchange correlation functional. The DFT-D3 calculations were considered to describe the weak dispersion forces proposed by Grimme [39,51]. Furthermore, Heyd–Scuseria–Ernzerhof (HSE06) hybrid functional calculations were conducted to obtain more accurate electronic properties [52]. Monkhorst–Pack k-point grids of $11 \times 11 \times 1$ and $17 \times 17 \times 1$ in the first Brillouin zone (BZ) were used for the relaxation and self-consistent calculations, respectively. The spin–orbit coupling (SOC) effect is not considered in this work, because it has a negligible effect on the electron band structure of the studied materials. The energy cut-off was set as 550 eV. To avoid interaction between nearby layers, the vacuum space adopted was 20 Å. The convergence values for force and energy were set within 0.01 eV Å^{-1} and 0.01 meV, respectively.

4. Conclusions

In summary, the mechanical, electronic and HER properties of the δ-IV–VI monolayers, namely, GeS, GeSe, SiS and SiSe, were systematically investigated using first-principles calculations. The strain–stress relationships of these δ-IV–VI monolayers present a novel toughness along the y direction, while the yield strength was calculated for the SiS and SiSe monolayers in the x direction. The GeS, GeSe, SiS and SiSe monolayers showed semiconductor characteristics with a bandgap larger than 1.23 eV for water splitting. For this application, an excellent carrier mobility was determined for all these δ-IV–VI monolayers; in particular, the GeSe monolayer demonstrates an ultrahigh electron mobility in the x direction of approximately 32,507 cm$^2 \cdot$V$^{-1} \cdot$s^{-1}. Furthermore, the Gibbs free energies of these GeS, GeSe, SiS and SiSe monolayers were obtained and imply their potential for usage as photocatalysts for water splitting.

Author Contributions: Conceptualization, Z.H. and K.R.; methodology, K.R. and L.W. (Liangmo Wang); software, L.W. (Liangmo Wang); validation, L.W. (Li Wang) and R.Z.; formal analysis, K.R.; investigation, Z.H.; resources, Z.H.; data curation, K.R.; writing—original draft preparation, Z.H.; writing—review and editing, K.R.; visualization, L.W. (Liangmo Wang); supervision, L.W. (Li Wang); project administration, L.W. (Li Wang); funding acquisition, L.W. (Li Wang), Z.H. and K.R. All authors have read and agreed to the published version of the manuscript.

Funding: We acknowledge the financial support of the Natural Science Foundation of Jiangsu (No. BK20220407) and Universities and Key Project of Natural Science Research of Anhui Provincial Department of Education (No. 2022AH052431) and the Open Fund Project of Maanshan Engineering Technology Research Center of Advanced Design for Automotive Stamping Dies (grant number: QMSG202105).

Institutional Review Board Statement: Not applicable.

Informed Consent Statement: Not applicable.

Data Availability Statement: The data presented in this study are available upon request from the corresponding author.

Acknowledgments: We gratefully acknowledge HZWTECH for providing computation facilities. We thank Jie Li from HZWTECH for help and discussions regarding this study.

Conflicts of Interest: The authors declare no conflict of interest.

Sample Availability: Samples of the compounds are available from the authors.

References

1. Geim, A.K.; Novoselov, K.S. The rise of graphene. *Nat. Mater.* **2007**, *6*, 183–191. [CrossRef]
2. Andrew, R.C.; Mapasha, R.E.; Ukpong, A.M.; Chetty, N. Mechanical properties of graphene and boronitrene. *Phys. Rev. B* **2012**, *85*, 125428. [CrossRef]

3. Zhang, H.; Chhowalla, M.; Liu, Z. 2D nanomaterials: Graphene and transition metal dichalcogenides. *Chem. Soc. Rev.* **2018**, *47*, 3015–3017. [CrossRef]
4. Wang, K.; Ren, K.; Hou, Y.; Cheng, Y.; Zhang, G. Physical insights into enhancing magnetic stability of 2D magnets. *J. Appl. Phys.* **2023**, *133*, 110902. [CrossRef]
5. Fei, H.; Dong, J.; Arellano-Jimenez, M.J.; Ye, G.; Kim, N.D.; Samuel, E.L.; Peng, Z.; Zhu, Z.; Qin, F.; Bao, J.; et al. Atomic cobalt on nitrogen-doped graphene for hydrogen generation. *Nat. Commun.* **2015**, *6*, 8668. [CrossRef]
6. Miro, P.; Audiffred, M.; Heine, T. An atlas of two-dimensional materials. *Chem. Soc. Rev.* **2014**, *43*, 6537–6554. [CrossRef]
7. Wang, G.; Zhi, Y.; Bo, M.; Xiao, S.; Li, Y.; Zhao, W.; Li, Y.; Li, Y.; He, Z. 2D Hexagonal Boron Nitride/Cadmium Sulfide Heterostructure as a Promising Water-Splitting Photocatalyst. *Phys. Status Solidi* **2020**, *257*, 1900431. [CrossRef]
8. Wang, G.; Zhang, L.; Li, Y.; Zhao, W.; Kuang, A.; Li, Y.; Xia, L.; Li, Y.; Xiao, S. Biaxial strain tunable photocatalytic properties of 2D ZnO/GeC heterostructure. *J. Phys. Phys. D Appl. Phys.* **2020**, *53*, 015104. [CrossRef]
9. Wang, G.; Gong, L.; Li, Z.; Wang, B.; Zhang, W.; Yuan, B.; Zhou, T.; Long, X.; Kuang, A. A two-dimensional CdO/CdS heterostructure used for visible light photocatalysis. *Phys. Chem. Chem. Phys.* **2020**, *22*, 9587–9592. [CrossRef] [PubMed]
10. Hou, Y.; Wei, Y.; Yang, D.; Wang, K.; Ren, K.; Zhang, G. Enhancing the Curie Temperature in $Cr_2Ge_2Te_6$ via Charge Doping: A First-Principles Study. *Molecules* **2023**, *28*, 3893. [CrossRef]
11. Ju, L.; Liu, P.; Yang, Y.; Shi, L.; Yang, G.; Sun, L. Tuning the photocatalytic water-splitting performance with the adjustment of diameter in an armchair WSSe nanotube. *J. Energy Chem.* **2021**, *61*, 228–235. [CrossRef]
12. Qin, H.; Zhang, G.; Ren, K.; Pei, Q.-X. Thermal Conductivities of PtX_2 (X = S, Se, and Te) Monolayers: A Comprehensive Molecular Dynamics Study. *J. Phys. Chem. C* **2023**, *127*, 8411–8417. [CrossRef]
13. Desai, S.B.; Seol, G.; Kang, J.S.; Fang, H.; Battaglia, C.; Kapadia, R.; Ager, J.W.; Guo, J.; Javey, A. Strain-induced indirect to direct bandgap transition in multilayer WSe_2. *Nano Lett.* **2014**, *14*, 4592–4597. [CrossRef]
14. Cai, Y.; Zhang, G.; Zhang, Y.W. Polarity-reversed robust carrier mobility in monolayer MoS_2 nanoribbons. *J. Am. Chem. Soc.* **2014**, *136*, 6269–6275. [CrossRef] [PubMed]
15. Ju, L.; Tang, X.; Li, X.; Liu, B.; Qiao, X.; Wang, Z.; Yin, H. NO_2 Physical-to-Chemical Adsorption Transition on Janus WSSe Monolayers Realized by Defect Introduction. *Molecules* **2023**, *28*, 1644. [CrossRef]
16. Jin, C.; Tang, X.; Tan, X.; Smith, S.C.; Dai, Y.; Kou, L. A Janus MoSSe monolayer: A superior and strain-sensitive gas sensing material. *J. Mater. Chem. A* **2019**, *7*, 1099–1106. [CrossRef]
17. Ren, K.; Wang, S.; Luo, Y.; Chou, J.-P.; Yu, J.; Tang, W.; Sun, M. High-efficiency photocatalyst for water splitting: A Janus MoSSe/XN (X = Ga, Al) van der Waals heterostructure. *J. Phys. Phys. D Appl. Phys.* **2020**, *53*, 185504. [CrossRef]
18. Ren, K.; Qin, H.; Liu, H.; Chen, Y.; Liu, X.; Zhang, G. Manipulating Interfacial Thermal Conduction of 2D Janus Heterostructure via a Thermo-Mechanical Coupling. *Adv. Funct. Mater.* **2022**, *32*, 2110846. [CrossRef]
19. Ju, L.; Tang, X.; Li, J.; Shi, L.; Yuan, D. Breaking the out-of-plane symmetry of Janus WSSe bilayer with chalcogen substitution for enhanced photocatalytic overall water-splitting. *Appl. Surf. Sci.* **2022**, *574*, 151692. [CrossRef]
20. Cui, Q.; Ren, K.; Zheng, R.; Zhang, Q.; Yu, L.; Li, J. Tunable thermal properties of the biphenylene and the lateral heterostructure formed with graphene: A molecular dynamics investigation. *Front. Phys.* **2022**, *10*, 1085367. [CrossRef]
21. Lu, N.; Zhuo, Z.; Guo, H.; Wu, P.; Fa, W.; Wu, X.; Zeng, X.C. CaP_3: A New Two-Dimensional Functional Material with Desirable Band Gap and Ultrahigh Carrier Mobility. *J. Phys. Chem. Lett.* **2018**, *9*, 1728–1733. [CrossRef]
22. Ren, K.; Yan, Y.; Zhang, Z.; Sun, M.; Schwingenschlögl, U. A family of Li_xB_y monolayers with a wide spectrum of potential applications. *Appl. Surf. Sci.* **2022**, *604*, 154317. [CrossRef]
23. Ren, K.; Shu, H.; Wang, K.; Qin, H. Two-dimensional MX_2Y_4 systems: Ultrahigh carrier transport and excellent hydrogen evolution reaction performances. *Phys. Chem. Chem. Phys.* **2023**, *25*, 4519–4527. [CrossRef] [PubMed]
24. Ju, L.; Tang, X.; Zhang, Y.; Li, X.; Cui, X.; Yang, G. Single Selenium Atomic Vacancy Enabled Efficient Visible-Light-Response Photocatalytic NO Reduction to NH(3) on Janus WSSe Monolayer. *Molecules* **2023**, *28*, 2959. [CrossRef]
25. Ouyang, Y.; Ling, C.; Chen, Q.; Wang, Z.; Shi, L.; Wang, J. Activating Inert Basal Planes of MoS_2 for Hydrogen Evolution Reaction through the Formation of Different Intrinsic Defects. *Chem. Mater.* **2016**, *28*, 4390–4396. [CrossRef]
26. Cai, Y.; Gao, J.; Chen, S.; Ke, Q.; Zhang, G.; Zhang, Y.-W. Design of Phosphorene for Hydrogen Evolution Performance Comparable to Platinum. *Chem. Mater.* **2019**, *31*, 8948–8956. [CrossRef]
27. Yan, P.; She, X.; Zhu, X.; Xu, L.; Qian, J.; Xia, J.; Zhang, J.; Xu, H.; Li, H.; Li, H. Efficient photocatalytic hydrogen evolution by engineering amino groups into ultrathin 2D graphitic carbon nitride. *Appl. Surf. Sci.* **2020**, *507*, 145085. [CrossRef]
28. Wang, X.; Liu, G.; Chen, Z.G.; Li, F.; Wang, L.; Lu, G.Q.; Cheng, H.M. Enhanced photocatalytic hydrogen evolution by prolonging the lifetime of carriers in ZnO/CdS heterostructures. *Chem. Commun.* **2009**, *23*, 3452–3454. [CrossRef]
29. Wang, B.-J.; Li, X.-H.; Cai, X.-L.; Yu, W.-Y.; Zhang, L.-W.; Zhao, R.-Q.; Ke, S.-H. Blue Phosphorus/$Mg(OH)_2$ van der Waals Heterostructures as Promising Visible-Light Photocatalysts for Water Splitting. *J. Phys. Chem. C* **2018**, *122*, 7075–7080. [CrossRef]
30. Ren, K.; Ma, X.; Liu, X.; Xu, Y.; Huo, W.; Li, W.; Zhang, G. Prediction of 2D IV–VI semiconductors: Auxetic materials with direct bandgap and strong optical absorption. *Nanoscale* **2022**, *14*, 8463–8473. [CrossRef]
31. Sun, M.; Schwingenschlögl, U. Structure Prototype Outperforming MXenes in Stability and Performance in Metal-Ion Batteries: A High Throughput Study. *Adv. Energy Mater.* **2021**, *11*, 2003633. [CrossRef]
32. Mao, Y.; Ben, J.; Yuan, J.; Zhong, J. Tuning the electronic property of two dimensional SiSe monolayer by in-plane strain. *Chem. Phys. Lett.* **2018**, *705*, 12–18. [CrossRef]

33. Van Thanh, V.; Van, N.D.; Truong, D.V.; Hung, N.T. Effects of strain and electric field on electronic and optical properties of monolayer γ-GeX (X = S, Se and Te). *Appl. Surf. Sci.* **2022**, *582*, 152321. [CrossRef]
34. Poudel, S.P.; Barraza-Lopez, S. Metastable piezoelectric group-IV monochalcogenide monolayers with a buckled honeycomb structure. *Phys. Rev. B* **2021**, *103*, 024107. [CrossRef]
35. Hu, Z.; Ding, Y.; Hu, X.; Zhou, W.; Yu, X.; Zhang, S. Recent progress in 2D group IV-IV monochalcogenides: Synthesis, properties and applications. *Nanotechnology* **2019**, *30*, 252001. [CrossRef] [PubMed]
36. Hu, P.; Wen, Z.; Wang, L.; Tan, P.; Xiao, K. Synthesis of few-layer GaSe nanosheets for high performance photodetectors. *ACS Nano* **2012**, *6*, 5988–5994. [CrossRef]
37. Wei, Y.; He, J.; Zhang, Q.; Liu, C.; Wang, A.; Li, H.; Zhai, T. Synthesis and investigation of layered GeS as a promising large capacity anode with low voltage and high efficiency in full-cell Li-ion batteries. *Mater. Chem. Front.* **2017**, *1*, 1607–1614. [CrossRef]
38. Shao, C.; Ren, K.; Huang, Z.; Yang, J.; Cui, Z. Two-Dimensional PtS2/MoTe2 van der Waals Heterostructure: An Efficient Potential Photocatalyst for Water Splitting. *Front. Chem.* **2022**, *10*, 847319. [CrossRef] [PubMed]
39. Van de Walle, C.G.; Martin, R.M. "Absolute"deformation potentials: Formulation and ab initio calculations for semiconductors. *Phys. Rev. Lett.* **1989**, *62*, 2028. [CrossRef]
40. Cai, Y.; Zhang, G.; Zhang, Y.-W. Layer-dependent band alignment and work function of few-layer phosphorene. *Sci. Rep.* **2014**, *4*, 6677. [CrossRef] [PubMed]
41. Zhang, L.; Tang, C.; Zhang, C.; Gu, Y.; Du, A. First-principles prediction of ferroelasticity tuned anisotropic auxeticity and carrier mobility in two-dimensional AgO. *J. Mater. Chem. C* **2021**, *9*, 3155–3160. [CrossRef]
42. Dai, J.; Zeng, X.C. Titanium trisulfide monolayer: Theoretical prediction of a new direct-gap semiconductor with high and anisotropic carrier mobility. *Angew. Chem.* **2015**, *54*, 7572–7576. [CrossRef] [PubMed]
43. Qiao, J.; Kong, X.; Hu, Z.X.; Yang, F.; Ji, W. High-mobility transport anisotropy and linear dichroism in few-layer black phosphorus. *Nat. Commun.* **2014**, *5*, 4475. [CrossRef] [PubMed]
44. Kumar, R.; Das, D.; Singh, A.K. C2N/WS2 van der Waals type-II heterostructure as a promising water splitting photocatalyst. *J. Catal.* **2018**, *359*, 143–150. [CrossRef]
45. Ren, K.; Luo, Y.; Wang, S.; Chou, J.-P.; Yu, J.; Tang, W.; Sun, M. A van der Waals Heterostructure Based on Graphene-like Gallium Nitride and Boron Selenide: A High-Efficiency Photocatalyst for Water Splitting. *ACS Omega* **2019**, *4*, 21689–21697. [CrossRef] [PubMed]
46. Zang, Y.; Wu, Q.; Du, W.; Dai, Y.; Huang, B.; Ma, Y. Activating electrocatalytic hydrogen evolution performance of two-dimensional MSi2N4 (M = Mo, W): A theoretical prediction. *Phys. Rev. Mater.* **2021**, *5*, 045801. [CrossRef]
47. Blöchl, P.E. Projector augmented-wave method. *Phys. Rev. B* **1994**, *50*, 17953. [CrossRef]
48. Kresse, G.; Furthmüller, J. Efficient iterative schemes for ab initio total-energy calculations using a plane-wave basis set. *Phys. Rev. B* **1996**, *54*, 11169. [CrossRef]
49. Perdew, J.P.; Burke, K. Ernzerhof, Generalized gradient approximation made simple. *Phys. Rev. Lett.* **1996**, *77*, 3865. [CrossRef]
50. Heyd, J.; Scuseria, G.E.; Ernzerhof, M. Hybrid functionals based on a screened Coulomb potential. *J. Chem. Phys.* **2003**, *118*, 8207–8215. [CrossRef]
51. Grimme, S.; Antony, J.; Ehrlich, S.; Krieg, H. A consistent and accurate ab initio parametrization of density functional dispersion correction (DFT-D) for the 94 elements H-Pu. *J. Chem. Phys.* **2010**, *132*, 154104. [CrossRef] [PubMed]
52. Heyd, J.; Peralta, J.E.; Scuseria, G.E.; Martin, R.L. Energy band gaps and lattice parameters evaluated with the Heyd-Scuseria-Ernzerhof screened hybrid functional. *J. Chem. Phys.* **2005**, *123*, 174101. [CrossRef] [PubMed]

Disclaimer/Publisher's Note: The statements, opinions and data contained in all publications are solely those of the individual author(s) and contributor(s) and not of MDPI and/or the editor(s). MDPI and/or the editor(s) disclaim responsibility for any injury to people or property resulting from any ideas, methods, instructions or products referred to in the content.

Article

Single Selenium Atomic Vacancy Enabled Efficient Visible-Light-Response Photocatalytic NO Reduction to NH₃ on Janus WSSe Monolayer

Lin Ju [1,*], Xiao Tang [2], Yixin Zhang [1], Xiaoxi Li [1], Xiangzhen Cui [1] and Gui Yang [3,*]

1 School of Physics and Electric Engineering, Anyang Normal University, Anyang 455000, China
2 Institute of Materials Physics and Chemistry, College of Science, Nanjing Forestry University, Nanjing 210037, China
3 School of Mechanical and Electrical Engineering, Chuzhou University, Chuzhou 239000, China
* Correspondence: julin@aynu.edu.cn (L.J.); fengmingchun@chzu.edu.cn (G.Y.)

Citation: Ju, L.; Tang, X.; Zhang, Y.; Li, X.; Cui, X.; Yang, G. Single Selenium Atomic Vacancy Enabled Efficient Visible-Light-Response Photocatalytic NO Reduction to NH₃ on Janus WSSe Monolayer. *Molecules* **2023**, *28*, 2959. https://doi.org/10.3390/molecules28072959

Academic Editor: Franca Morazzoni

Received: 14 February 2023
Revised: 22 March 2023
Accepted: 24 March 2023
Published: 26 March 2023

Copyright: © 2023 by the authors. Licensee MDPI, Basel, Switzerland. This article is an open access article distributed under the terms and conditions of the Creative Commons Attribution (CC BY) license (https://creativecommons.org/licenses/by/4.0/).

Abstract: The NO reduction reaction (NORR) toward NH₃ is simultaneously emerging for both detrimental NO elimination and valuable NH₃ synthesis. An efficient NORR generally requires a high degree of activation of the NO gas molecule from the catalyst, which calls for a powerful chemisorption. In this work, by means of first-principles calculations, we discovered that the NO gas molecule over the Janus WSSe monolayer might undergo a physical-to-chemical adsorption transition when Se vacancy is introduced. If the Se vacancy is able to work as the optimum adsorption site, then the interface's transferred electron amounts are considerably increased, resulting in a clear electronic orbital hybridization between the adsorbate and substrate, promising excellent activity and selectivity for NORR. Additionally, the NN bond coupling and *N diffusion of NO molecules can be effectively suppressed by the confined space of Se vacancy defects, which enables the active site to have the superior NORR selectivity in the NH₃ synthesis. Moreover, the photocatalytic NO-to-NH₃ reaction is able to occur spontaneously under the potentials solely supplied by the photo-generated electrons. Our findings uncover a promising approach to derive high-efficiency photocatalysts for NO-to-NH₃ conversion.

Keywords: NO reduction; WSSe monolayer; photocatalysis; density functional theory

1. Introduction

A heightened consciousness of environmental and health issues has prompted significant endeavors to discover efficient and cost-effective technologies to detect, regulate, and remove a wide range of air pollutants, for example, nitric oxide (NO$_x$), particulate matter (PM), and sulfur oxide (SO$_x$). In this respect, NO, which is mainly emitted from the combustion of fossil fuels in stationary thermal power plants and internal combustion engines [1], is regarded as an essential threat to both human health and the global climate, given that it is a major factor in the formation of harmful photochemical smog, haze, and acid rain, etc. [2,3]. It has been reported that several methods, involving physical/chemical adsorption [4], heterogeneous catalytic reduction [5,6], and oxidation [7], have shown high efficiency in the selective sequestration and conversion of NO. However, such approaches have always been worked on especially for the treatment of NO in relatively high concentrations in the atmosphere, and both the capital and energy become unaffordable in the removal of NO at the ppb level. The development of an approach with the following characteristics is highly desirable but still challenging for practical ppb-level NO treating, i.e., significant NO conversion efficiency at room temperature, reliable performance in large-scale gas purification, and low-cost energy investment.

Moreover, in the field of NO conversion, where the N=O bond (204 KJ/mol) is more easily activated than the N≡N bond (941 KJ/mol), NH₃ produced by electrocatalytic NO reduction reaction (NORR), as an attractive candidate for the traditional nitrogen reduction reaction (NRR), has recently been realized with many electrocatalysts [8–10]. Compared with the electrocatalytic process, the photocatalytic one is normally more attractive, since it is green, sustainable, and it inherits the advantages of natural photosynthesis. Though there are some achievements for the photocatalytic NRR that uses water as a proton source and reaction solvent [11–17], developing a stable and highly efficient photocatalyst for solar-driven NORR to NH_3, to our knowledge, is still rarely reported.

Janus 2D transition metal dichalcogenides (TMD), an emerging 2D material that refers to layered materials with diverse surfaces, has recently sparked a lot of research interest in applications for photocatalytic energy conversion [18–21]. It is predicted that in Janus 2D TMD materials, the out-of-plane structural asymmetry-caused intrinsic dipole can manage the incompatible demands of the band gap between the high light-utilization rate and the capable redox capacity. Specifically, a sufficiently narrow band gap is usually required for high light utilization; however, a large band gap ($\geq$1.23 eV) is usually required for a sufficient redox capability. With respect to two-dimensional polar photocatalysts, Yang et al. proposed that due to the presence of polarization, the top of the valence band and the bottom of the conduction band should be distributed on opposite sides, bringing about a potential difference that would improve the redox ability of the photoexcited carriers and reduce the requirement for a band gap [22]. Therefore, the Janus TMD materials are predicted to have a better photocatalytic property than the symmetrical traditional TMD materials [20,23]. Through replacing S atoms of WS_2 with Se atoms with pulsed laser ablation plasmas [24] and hotting WSe_2 and WS_2 mixed powders under 1000 °C [25], the Janus WSSe monolayer, a classical Janus 2D TMD, has been successfully produced. Recently, Janus WSSe monolayers have also been reported to have significant applicability potential for photocatalytic overall water-splitting due to the excellent optical absorption, adequate redox capability, and high carrier separation [18].

Here, through density functional theory (DFT) calculations, we investigated NO adsorption upon 2H phase Janus WSSe monolayers, with and without manufactured Se-vacancy defects. The metallic 1T phase is not considered in our work, because it is typically unstable under ambient conditions [26]. To comprehend the coupling effect between NO and a substrate, a systematic explanation of the adsorption energy, charge density differential (CDD), and density of states (DOS) has been provided. We discovered that adding Se vacancy could lead to a shift in the physical-to-chemical nature of NO gas adsorption on the Janus WSSe monolayer, which could effectively activate NO gas molecules, thus making the NORR possible. Then, we investigated the photocatalytic NORR on the defective Janus WSSe monolayer, by estimating the optical absorption, redox capacity, and the driving force of the photo-exited electron for NORR. At last, we discuss the competition between the NORR and hydrogen evolution reaction (HER). Our results reveal that the defective WSSe with the outstanding photocatalytic NORR performance could be used as a novel platform for NO conversion.

2. Results and Discussion

2.1. NO Physical Absorption upon a Pristine Janus WSSe Monolayer

2.1.1. Adsorbing Site Selection and E_{ads}

A W layer is sandwiched between the S and Se layers to form the Janus WSSe single layer. Like WSe_2 or WS_2, the matrix materials, the Janus WSSe monolayer has a honeycomb structure [27]. According to calculations, Janus WSSes lattice constant is 3.26 Å, which is in the middle of the range between WS_2 and WSe_2, which are its parent materials. It is highly desirable to investigate whether the vertical intrinsic dipole that is brought on by the asymmetric structure will enhance the gas sensing properties of Janus WSSe, similar to how it did in the case of Janus MoSSe [28]. In this work, geometric properties of NO adsorption on both sides of Janus WSSe were initially taken into consideration, as seen in Figure 1a,c. Every adsorption situation involves placing one NO molecule upon a WSSe monolayer's 4 × 4 supercell whereas the system is fully relaxed. In addition, several potential adsorption sites have been taken into account, i.e., the top sites over the W-Se/(W-S) bond (named **Bond**), S/Se/W atom (named **W/Se/S**), and hexagon's center (named **Center**).

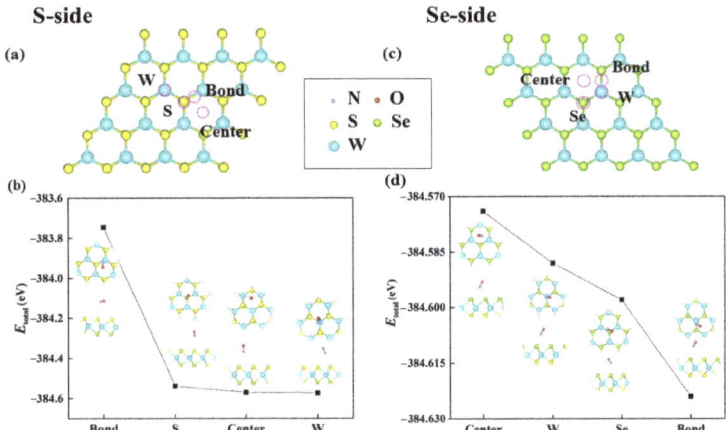

Figure 1. (a) S-side and (c) Se-side possible adsorption sites (symbolized in purple circles) considered for the case of pristine WSSe. The E_{total} of (b) S-side and (d) Se-side NO adsorption systems with various adsorption sites. The N, O, S, Se, and W atoms are represented with the purple, red, yellow, green, and blue balls, respectively, and this color scheme is also used in Figures 2, 4, 5, 7 and 8.

In accordance with Equation (1), it is observed that E_{total} dominates E_{ads}, since E_{sub} and E_{gas} are invariant at the various adsorption sites. Herewith, to explore the most stable adsorption configuration, we calculated the E_{total} of the NO adsorption in both the S-side and Se-side patterns by considering various adsorption sites. For the NO adsorption on the S-side, as shown in Figure 1b, we found that E_{total} achieves the minimum (-384.57 eV) if it lies on the **W** site, denoting the most stable adsorption conformation. For the Se-side NO adsorption, once the molecule was placed over the **Bond** site, the system had the lowest E_{total} (-384.62 eV), implying the steadiest adsorption site, as shown in Figure 1d. Additionally, the Se-side steadiest adsorption system E_{total} is lower than the S-side one by 0.05 eV, leading to the conclusion that NO gas molecules are more likely to adsorb in the Se plane. Consequently, we chose the Se-side NO adsorption on the **Bond** site to stand for the case of NO gas molecule adsorbing on the pristine Janus WSSe monolayer. The absolute value of E_{ads} for a physisorption is often less than 1 eV [29–32]. Therefore, the adsorption for the configuration is most likely a physisorption with the E_{ads} equal to -0.21 eV. This will be discussed in the following section for further investigations of the physisorption.

2.1.2. Adsorption Mechanism

The NO molecule kept parallel to the surface of the substrate after adsorption at a vertical distance away of 2.72 Å as its nitrogen atoms tended to the surface, as seen in Figure 2a. Additionally, the smallest distance from the NO molecule's N atom to its nearest Se atom reaches 3.53 Å, which is significantly longer than the Se-N bond's length of 1.81Å. Moreover, as shown in Figure 2b, the CDD results indicate that the charge redistribution mainly takes place at the NO gas molecule, and only very rare electrons (merely 0.052 e) migrate from the substrate to the NO molecule, resulting in the weak interaction between them.

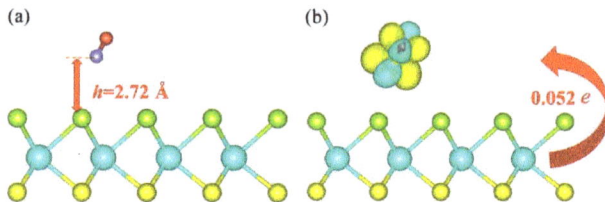

Figure 2. The optimized structure in a side view (**a**) and CDD image (**b**) for adsorption system, composed of pristine Janus WSSe monolayer and NO gas molecule. The h in red represents the adsorption distance from NO to pristine Janus WSSe monolayer. Charge deposition (exhaustion) is indicated by yellow (cyan) areas. The value of the isosurface is $0.002\ e\ Å^{-3}$, and that red arrow indicates the charge transfer direction.

This adsorption configuration's pertinent DOS has been computed. Figure 3a–c shows how little the gas molecule as well as the monolayer were altered after adsorption, with respect to DOS. This is consistent with the minute interface transfer electron, which suggests that neither the electronic properties of WSSe nor NO have changed noticeably. The very weak connection seen between the WSSe monolayer and NO is shown by their poor orbital hybridization, which is consistent with the previous statement. Additionally, as shown in Figure 3d, the Se p orbital from the Se atom in the WSSe monolayer, most near the NO gas molecule, as well as the N p orbital from the N atom in the NO gas molecule, are independent of one another. Based on the investigation above, the NO adsorption over the pristine WSSe should be physisorption.

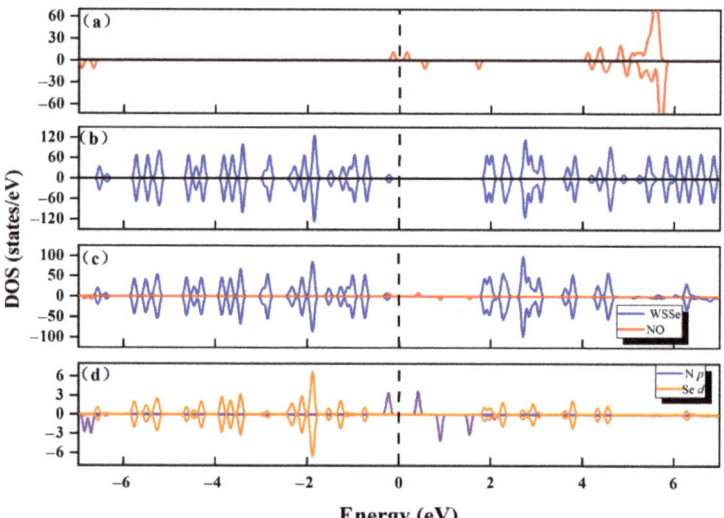

Figure 3. Total DOS of (**a**) a free NO molecule, as well as (**b**) a pure pristine WSSe. (**c**) The adsorption system's partial DOS. Dark blue is used to represent the WSSe component while red is used to represent the NO portion. (**d**) N p orbitals (marked in purple) of the adsorbed NO and Se p orbitals (marked in orange) of the Se atoms most near adsorbed NO molecule. The Fermi level is indicated by the perpendicular dashed line.

2.2. NO Chemisorption and Reduction Reaction over Defective Janus WSSe Monolayer

NO physisorption on pristine WSSe is suitable for use in gas collection systems. Yet, the need for NO chemisorption is greater when it comes to treating gases or accelerating chemical reactions, which calls for a substrate with a stronger adsorption capability. Based on earlier pertinent findings, it is found that adding a few vacancy defects might affect the electrical property and hence significantly increase the stability of specific geometric formations [33,34]. Therefore, we create vacancy defects in the Janus WSSe monolayer in an effort to increase NO gas molecule adsorption. Here, we concentrate on the single Se vacancy defect for the following three reasons: (I) The Se vacancy is easier to form than other kinds of vacancy defects at the Janus TMD monolayer due to its relatively lower formation energy [35]. (II) As previously mentioned, NO gas molecules tend to adsorb on the Se-side of the pristine monolayer. (III) Photo-reduction has been theoretically demonstrated to take place on the Se-side of the pristine monolayer [18], which shows a potential for the NORR to NH_3.

2.2.1. Adsorbing Site Selection and E_{ads}

As depicted in Figure 4a, for the defective WSSe monolayer, five possible adsorption sites were taken into consideration. They are the **Center** (the top site above the center of the hexagon), **W** and **Se** (the top of the W and Se atoms, separately), **Bond** (the top site above the W-Se bond), and **Vacancy** (the top site above the Se vacancy defect) adsorption sites. The adsorption system E_{total} was employed to capture the most likely adsorption morphology, analogous to the pristine monolayer situation. The E_{total} was minimized when NO was adsorbed on the **Vacancy** site (see Figure 4b), so the **Vacancy** site is the most likely adsorption site in this case. The E_{ads} under the condition is -2.92 eV, which represents an order of magnitude that is more negative than that on pristine WSSe (-0.21 eV). It is clear that the introduction of Se vacancies resulted in an effective enhancement of the NO adsorption. From the anomalously negative E_{ads}, it can be tentatively determined that this NO adsorption on defective WSSe is chemisorption. We explore this issue in more depth in the next section.

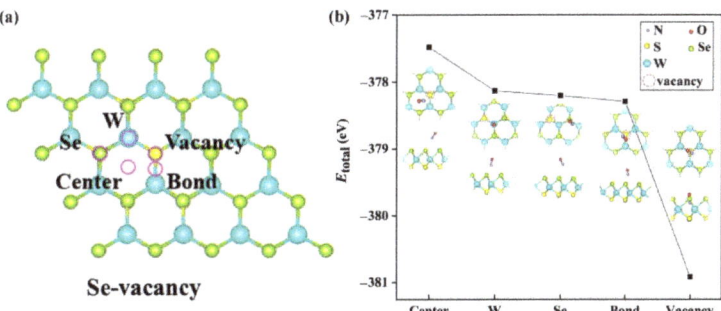

Figure 4. (a) Possible adsorption sites (symbolized in purple circles) considered for the case of defective WSSe. (b) The E_{total} of this NO adsorption system with various adsorption sites.

2.2.2. Adsorption Mechanism

The nitrogen atom in the N-O bond of the NO gas molecule takes a nearly vertical orientation, as seen in Figure 5a, pointing to the surface of the monolayer. At the surface of the monolayer, the nitrogen atom connects with the three tungsten atoms that are next to it. As a result, the adsorption is unquestionably chemisorption, which is consistent with that outcome produced by its adsorption energy as stated before. Additionally, we evaluated the N-O bond length for quantitatively analyzing how the morphology of NO changed pre and post adsorption. Before adsorption, it is 1.17 Å, and as shown in Figure 5b, it stretches to 2.13 Å post adsorption, indicating electron redistribution in NO through the adsorption process. A large number of electrons (1.04 e) move to the adsorbate from the damaged Janus WSSe layer, as can be observed in Figure 5c, where there are notable charge redistributions in the adsorption system. For gas sensors, resistivity fluctuation is typically brought on by adsorption-induced charge transfer, which is a crucial indicator of sensing merits and could be determined by experiments [36,37].

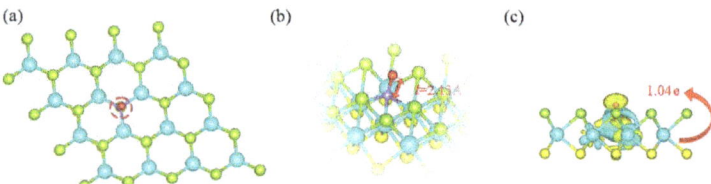

Figure 5. (a) Top, (b) side, and (c) CDD images of the optimal structures for the adsorption system under the defective WSSe case. The *l* in red stands for N-O bond length. Charge accumulation (depletion) is indicated by areas that are yellow (cyan). The value of the isosurface is 0.002 e Å^{-3}. The red number in CDD image indicates how much charge transferred from the substrate to the molecule.

We compute the pertinent DOS and display them in Figure 6 to obtain a greater understanding of the electronic characteristics for this chemisorption system. The two parts of the chemisorption system have a strong electronic orbital hybridization (see Figure 6b). This demonstrates that they interact strongly, which accounts for the observation that NO was closely bound to the substrate. Additionally, the coupling between the N p orbital from NO and W d orbitals of the W atoms, which bond to the N atom of NO, contributes significantly to the interaction (see Figure 6c). The comparison of the DOS of NO gas molecules between pre and post adsorption (see Figures 3a and S1) shows that the DOS is significantly delocalized after adsorption, indicating a sharp electron redistribution in NO, which is responsible for the visual N-O bond shift. From these results, we further demonstrate that the NO adsorption over the defective WSSe monolayer is chemisorption. Additionally, adding Se vacancies into Janus WSSe can wondrously trigger the NO physisorption-to-chemisorption transition.

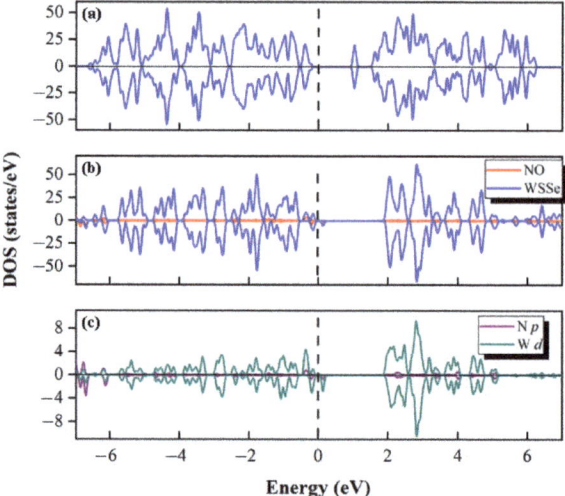

Figure 6. (a) Total DOS of the defective WSSe without NO adsorption. (b) The adsorption system's partial DOS. Dark blue indicates the WSSe component, whereas red indicates the NO portion (the enlarged view is shown in Figure S1). (c) The N p orbitals from the adsorbed NO gas molecule (marked in dark purple) and the W d (marked in dark green) orbitals of these three W atoms that are attaching to the N atom from NO. The Fermi level is indicated by the vertical dashed line.

2.2.3. Photocatalytic NORR

The obvious N-O bond elongation of the NO gas molecule caused by adsorption indicates that this molecule is activated, thus making the further NORR possible. Since the defective Janus WSSe monolayer is semiconductor with a bandgap of 1.22 eV (see Figure 6a), which is not suitable to act as an electrocatalysts, we study the photocatalytic NORR on the defective WSSe next.

The band edges of a semiconductor must line up with the potentials of the redox half-reactions in order for it to be active for the NO photo-reduction. Whether the photocatalytic NORR can proceed spontaneously depends directly on the strength of the external potentials that are provided by the photo-generated carriers [38]. The energy difference between the electron acceptor states and the hydrogen reduction potential (H^+/H_2) and the potential of photogenerated electrons for NORR (U_e) (Figure 7a) has been reported to be -1.11 V for the Janus WSSe monolayer at pH = 0 [18], which is significantly more negative than the theoretical potential of NORR (0.77 V vs. RHE [39]). A good resistance to photoinduced corrosion is facilitated by high U_e, which denotes the fact that photogenerated electrons of the Janus WSSe monolayer would prefer to be transferred to react with H^+ rather than with themselves [38,40].

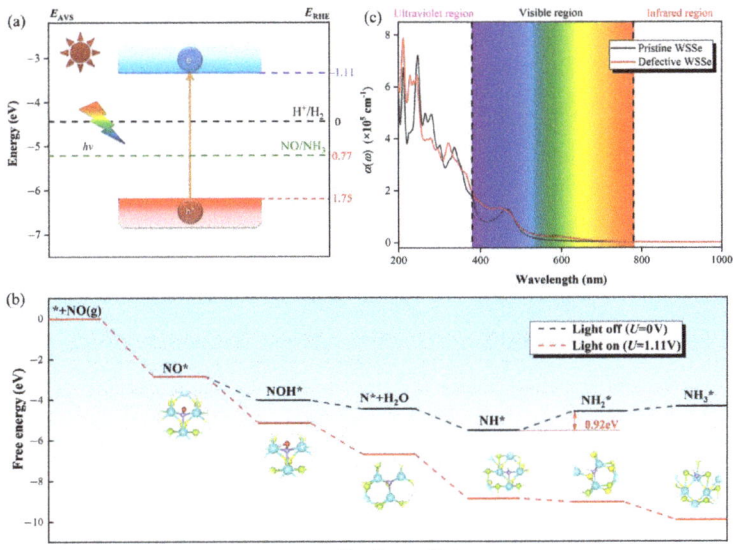

Figure 7. (a) A schematic representation of the Janus WSSe monolayer's band edge positions in relation to the reversible hydrogen electrode (RHE) at pH = 0. * stands for the adsorption site at the surface of catalyst. Relative energy levels to the absolute vacuum scale (AVS) and RHE are represented by E_{AVS} and E_{RHE}. (b) Gibbs free energy diagrams of NORR to NH_3 on defective Janus WSSe monolayer under $U = 0$ and $U = 1.11$ V. The applied potential that a photo-excited electron provides is $U = 1.11$ V. (c) The Janus WSSe monolayer's optical absorbance in both pristine and defective states.

There are five proton-coupled electron transfer steps during the NORR to NH_3 process (NO + $5H^+$ + $5e^-$ →NH_3 + H_2O). The free energy diagram as well as the related intermediate products for the NORR to NH_3 on the defective Janus WSSe monolayer are given in Figure 7b. The most favorable path is NO*→NOH*→N*→NH*→NH_2*→NH_3*. We can see that the electrocatalytic steps, including NO*→NOH*, NOH*→N*, and *N→*NH, are exothermic by -1.21, -0.44, and -1.05 eV, respectively. The third electrocatalytic step, i.e., N*+ e^- + H^+ → NH*, means that one H atom adsorbs on the N* to form NH*. In order to explore the ease of the NH* formation, we add the detailed analysis on the interaction between the H atom and N atom in the NH* based on the partial DOS. As shown in Figure S3, there is an obvious hybridization between N p and H s orbitals near the Fermi level, causing a strong attraction to each other. Therefore, the H atom could easily adsorb on the N*, making the reaction of NH* formation exothermic. Moreover, the exothermic reaction of NH* formation from hydrogenating N* on various electro-/photo-catalysts has been reported [6,41–43]. The other electrocatalytic steps, i.e., NH*→NH$_2$* and NH$_2$*→NH$_3$*, are endothermic separately, with free energy uphills of 0.92 and 0.24 eV. Excitingly, all NORR intermediate processes become exothermic when taking into account the external potential provided by photo-excited electrons (U = 1.11 V), demonstrating the spontaneous NORR with lighting (red line in Figure 7b).

An efficient photocatalyst must have a high photoconversion efficiency in order to start the photocatalytic conversion of NO to NH$_3$. Due to the narrowed direct band gap (see Figure S2), notably, Se vacancy introduction on the Janus WSSe monolayer leads to a redshift of the initial optical absorption peak (at 600.45 nm, red line), which is relative to the baseline value from the pristine Janus WSSe monolayer (at 466.28 nm, black line); therefore, it expands the optical absorption into visible regions, as shown in Figure 7c. Moreover, notably, the biggest absorption peak of the defective Janus WSSe monolayer among visible region, reaches up to 1.35×10^5 cm^{-1} (at 444.42 nm, red line), exceeding the one of pristine Janus WSSe monolayer (1.30×10^5 cm^{-1} at 466.28 nm, black line), which is comparable to some other photocatalysts, such as MoSSe–GaN (2.74×10^5 cm^{-1} at 425 nm) [44], MoSSe–AlN (3.95×10^5 cm^{-1} at 412 nm) [44], and graphene–MoSSe (about 4.00×10^5 cm^{-1} at 500 nm) [45]. The broadened optical absorption region and elevated optical absorption peak reveal that photons within a wider energy range can be utilized by bringing in the Se vacancy defect in the Janus WSSe monolayer.

2.2.4. Selectivity for NORR vs. HER

By depleting proton–electron pairs out of an electrolyte solution, the substantial competitive side reaction known as the hydrogen evolution reaction (HER) may drastically reduce the faradaic efficiency of NORR [46,47]. According to the Brønsted–Evans–Polanyi relation [48,49], lower ΔG reactions have lower reaction barriers and are therefore kinetically more preferred. Hereby, as shown in Figure 8a, we calculated the Gibbs free energy difference of H* formation (ΔG_{H^*}), and compared it with the one of NO* formation (ΔG_{NO^*}). The **Vacancy** site is the most feasible adsorption site for single H atom in defective Janus WSSe (more details of screening process can be found in the supporting information). As displayed in Figure 8b, ΔG_{NO^*} (−2.83 eV) is much lower than ΔG_{H^*} (0.73 eV), indicating that the active sites in the defective Janus WSSe monolayer will be preferentially occupied by *NO. According to the previous method used to judge the selectivity between HER and NORR [50], we could draw a conclusion that, NORR is highly preferred over HER.

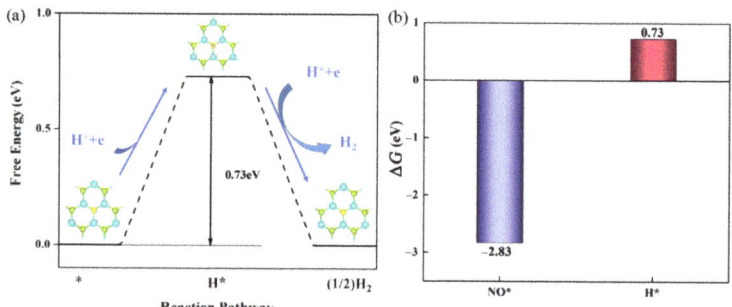

Figure 8. (a) Gibbs free energy diagram of HER on defective Janus WSSe monolayer. (b) ΔG_{NO^*} vs. ΔG_{H^*} of defective Janus WSSe monolayer.

2.2.5. Selectivity for NO-to-NH₃ Pathway vs. Other NORR Pathways

Besides HER, some other side reactions, such as the formation of N_2O_2 and N_2, perhaps restrain the production of NH_3 as well, so the selectivity of the reaction pathways for NORR should be considered. As mentioned before, due to spatial confinement, NO molecules can only ever assume the end-on orientation due to the N atom coupled with the exposed metal sites when adsorbing on the defective Janus WSSe monolayer. The reaction can only take place through the distal channel according to this NO adsorption model [51,52]. Most crucially, the Se vacancy defect's constrained space will successfully block the approach of two NO molecules, preventing the formation of N_2O_2, and N atoms can be firmly bound by the under-coordinated active sites in the vacancy to obstruct *N diffusion. Hence, N_2 production is excluded as a result of spatial constriction. Hence, there is a high selectivity of the NO-to-NH₃ reaction pathway guaranteed at a defective WSSe.

3. Materials and Methods

In this study, all the DFT simulations are operated with the Vienna Ab initio Simulation Program software package (Hanger team, University of Vienna, version 5.3) [53,54]. The exchange–correlation energy was simulated using the generalized gradient approximation of Perdew–Burke–Ernzerhof. We utilize the zero-damped DFT-D3 approach suggested by Grimme [55] to characterize the van der Waals (vdW) force. The plane wave basis set's cutoff energy was determined to be 500 eV. With a fixed lattice constant, all the internal coordinates were let to relax throughout optimization. As NO is a paramagnetic molecule, spin polarization is used when calculating the NO adsorption [56]. The computational model is built by one NO gas molecule adsorbing on a 4 × 4 supercell of pristine/defective Janus WSSe monolayer. Employing a 2 × 2 × 1 K point sampling, the Brillouin zone was sampled for integration using the Monkhorst-Pack technique [57] for structural optimization and electronic properties computations. To minimize the impact of interlayer contact, a 30 Å space was given down the direction that is normal to the plane. Moreover, the calculation of Gibbs free energy change for NORR is operated with the computational hydrogen electrode (CHE) model [58], and the solvent effect is considered with the implicit solvent model implemented in VASPsol [59,60]. More simulation details of the Gibbs free energy can be found in the supporting information.

The following formula is used to calculate NOs adsorption energy (E_{ads}) on both the damaged and unaltered WSSe monolayer [61,62],

$$E_{ads} = E_{total} - E_{sub} - E_{gas} \qquad (1)$$

where E_{sub} and E_{gas} separately are the clean substrate (pristine/defective Janus WSSe monolayer) and the sole NO molecule total energies, while E_{total} is the adsorption system total energy. An exothermic adsorption is indicated by a negative value for E_{ads}. The strength of the gas adsorption increases as E_{ads} is more negative.

The following equation was used to carry out the plane-integrated CDD,

$$\Delta\rho = \rho_{total} - \rho_{sub} - \rho_{gas} \qquad (2)$$

where ρ_{gas} and ρ_{sub} independently represent the charge density of the NO molecule and substrate, meanwhile, ρ_{total} is the adsorption system charge density.

The absorption coefficient $a(\omega)$ to assess the ability of sunlight harvesting is calculated following the formula below [63],

$$a(\omega) = \sqrt{2}\frac{\omega}{c}\left(\sqrt{\varepsilon_1(\omega)^2 + \varepsilon_2(\omega)^2} - \varepsilon_1(\omega)\right)^{\frac{1}{2}} \qquad (3)$$

where the real and imaginary components of a frequency-dependent dielectric function are denoted by ε_1 and ε_2, respectively, while the vacuum speed of light is denoted by c.

4. Conclusions

In our work, the NO adsorption on both pristine and defective WSSe monolayers has been theoretically investigated. On the pristine WSSe monolayer, the NO adsorption is physisorption based on minor adsorption energy, a large adsorption distance, and feeble electron orbital hybridization. By adding Se vacancies to WSSe, it is possible to convert the NO physisorption into chemisorption by significantly increasing the amount of interfacially transferred electrons and inducing significant electronic orbital coupling between the two components of the adsorption system. The powerful NO chemisorption gives defective WSSe high activity and selectivity for NORR. The active site has strong NORR selectivity for NH_3 production because the limited area of the Se vacancy defect may efficiently hinder the N-N bond coupling of NO molecules and the *N diffusion. Moreover, the potential provided by photogenerated electrons in the defective Janus WSSe monolayer is sufficient to drive a spontaneous NORR to NH_3. Our findings suggest an energy-saving and environmentally friendly strategy for direct NO-to-NH_3 conversion, which is anticipated to spur greater investigation into photocatalysts for NO-to-NH_3 conversion.

Supplementary Materials: The following supporting information can be downloaded at: https://www.mdpi.com/article/10.3390/molecules28072959/s1, Figure S1: The enlarged view for the partial density of states of NO portion from the adsorption system; Figure S2: The band structures of the pristine and defective Janus WSSe monolayers; Figure S3: The N p orbitals and the H s orbitals of intermediate NH*; Figure S4: Top view of the optimal structures for H* with H atom on **Center** and **Se** sites in the defective WSSe monolayer; Table S1: The total energy of H* with H atom on different deposition sites; Screening adsorption site for single H atom in defective Janus WSSe; Free energy difference in NORR.

Author Contributions: Supervision, L.J.; project administration, L.J. and G.Y.; Software, L.J. and Y.Z.; data curation, Y.Z. and X.L.; formal analysis, X.T. and X.C.; funding acquisition, L.J. and G.Y.; investigation, X.T., X.C. and L.J.; Writing—original draft, Y.Z., X.C., X.T. and L.J. All authors have read and agreed to the published version of the manuscript.

Funding: Our work is funded by the National Natural Science Foundation of China (Grant No. U20041103), the Natural Science Foundation of Henan Province (Grant No. 232300420128), the Henan Scientific Research Fund for Returned Scholars, the College Students Innovation Fund of Anyang Normal University (Grant No. 202210479049), the National College Students Innovation and Entrepreneurship Training Program (Grant No. 202210479032), the Open Project of Key Laboratory of Functional Materials and Devices for Informatics of Anhui Higher Education Institutes (Grant No. FSKFKT002), the Scientific and Technological Project of Anyang City (Grant Nos. 2020022, 2021C01GX014), and the Key Scientific and Technological Projects in Anyang City (Grant No. 2022C01GX019).

Institutional Review Board Statement: Not applicable.

Informed Consent Statement: Not applicable.

Data Availability Statement: The data presented in this study are available in Supplementary Materials.

Conflicts of Interest: The authors declare no conflict of interest.

References

1. Zhang, L.; Liang, J.; Wang, Y.; Mou, T.; Lin, Y.; Yue, L.; Li, T.; Liu, Q.; Luo, Y.; Li, N.; et al. High-Performance Electrochemical NO Reduction into NH_3 by MoS_2 Nanosheet. *Angew. Chem. Int. Edit.* **2021**, *60*, 25263–25268. [CrossRef] [PubMed]
2. Akimoto, H. Global Air Quality and Pollution. *Science* **2003**, *302*, 1716–1719. [CrossRef]
3. Fiore, A.M.; Naik, V.; Spracklen, D.V.; Steiner, A.; Unger, N.; Prather, M.; Bergmann, D.; Cameron-Smith, P.J.; Cionni, I.; Collins, W.J.; et al. Global air quality and climate. *Chem. Soc. Rev.* **2012**, *41*, 6663–6683. [CrossRef] [PubMed]
4. Rezaei, F.; Rownaghi, A.A.; Monjezi, S.; Lively, R.P.; Jones, C.W. SO_x/NO_x Removal from Flue Gas Streams by Solid Adsorbents: A Review of Current Challenges and Future Directions. *Energy Fuels* **2015**, *29*, 5467–5486. [CrossRef]
5. Xiong, S.; Weng, J.; Liao, Y.; Li, B.; Zou, S.; Geng, Y.; Xiao, X.; Huang, N.; Yang, S. Alkali Metal Deactivation on the Low Temperature Selective Catalytic Reduction of NO_x with NH_3 over MnO_x-CeO_2: A Mechanism Study. *J. Phys. Chem. C* **2016**, *120*, 15299–15309. [CrossRef]

6. He, B.; Lv, P.; Wu, D.; Li, X.; Zhu, R.; Chu, K.; Ma, D.; Jia, Y. Confinement catalysis of a single atomic vacancy assisted by aliovalent ion doping enabled efficient NO electroreduction to NH_3. *J. Mater. Chem. A* **2022**, *10*, 18690–18700. [CrossRef]
7. Wang, A.; Guo, Y.; Gao, F.; Peden, C.H.F. Ambient-temperature NO oxidation over amorphous CrO_x-ZrO_2 mixed oxide catalysts: Significant promoting effect of ZrO_2. *Appl. Catal. B-Environ.* **2017**, *202*, 706–714. [CrossRef]
8. Wang, J.; Feng, T.; Chen, J.; Ramalingam, V.; Li, Z.; Kabtamu, D.M.; He, J.-H.; Fang, X. Electrocatalytic nitrate/nitrite reduction to ammonia synthesis using metal nanocatalysts and bio-inspired metalloenzymes. *Nano Energy* **2021**, *86*, 106088. [CrossRef]
9. Mou, T.; Long, J.; Frauenheim, T.; Xiao, J. Advances in Electrochemical Ammonia Synthesis Beyond the Use of Nitrogen Gas as a Source. *ChemPlusChem* **2021**, *86*, 1211–1224. [CrossRef]
10. Wang, Y.; Wang, C.; Li, M.; Yu, Y.; Zhang, B. Nitrate electroreduction: Mechanism insight, in situ characterization, performance evaluation, and challenges. *Chem. Soc. Rev.* **2021**, *50*, 6720–6733. [CrossRef]
11. Liu, Y.; Ye, X.; Li, R.; Tao, Y.; Zhang, C.; Lian, Z.; Zhang, D.; Li, G. Boosting the photocatalytic nitrogen reduction to ammonia through adsorption-plasmonic synergistic effects. *Chin. Chem. Lett.* **2022**, *33*, 5162–5168. [CrossRef]
12. Ling, C.; Niu, X.; Li, Q.; Du, A.; Wang, J. Metal-Free Single Atom Catalyst for N_2 Fixation Driven by Visible Light. *J. Am. Chem. Soc.* **2018**, *140*, 14161–14168. [CrossRef] [PubMed]
13. Di, J.; Xia, J.; Chisholm, M.F.; Zhong, J.; Chen, C.; Cao, X.; Dong, F.; Chi, Z.; Chen, H.; Weng, Y.X.; et al. Defect-Tailoring Mediated Electron-Hole Separation in Single-Unit-Cell Bi_3O_4Br Nanosheets for Boosting Photocatalytic Hydrogen Evolution and Nitrogen Fixation. *Adv. Mater.* **2019**, *31*, e1807576. [CrossRef] [PubMed]
14. Yuan, J.; Yi, X.; Tang, Y.; Liu, M.; Liu, C. Efficient Photocatalytic Nitrogen Fixation: Enhanced Polarization, Activation, and Cleavage by Asymmetrical Electron Donation to N≡N Bond. *Adv. Funct. Mater.* **2019**, *30*, 1906983. [CrossRef]
15. Hou, T.; Peng, H.; Xin, Y.; Wang, S.; Zhu, W.; Chen, L.; Yao, Y.; Zhang, W.; Liang, S.; Wang, L. Fe Single-Atom Catalyst for Visible-Light-Driven Photofixation of Nitrogen Sensitized by Triphenylphosphine and Sodium Iodide. *ACS Catal.* **2020**, *10*, 5502–5510. [CrossRef]
16. Zhao, Z.; Choi, C.; Hong, S.; Shen, H.; Yan, C.; Masa, J.; Jung, Y.; Qiu, J.; Sun, Z. Surface-engineered oxidized two-dimensional Sb for efficient visible light-driven N_2 fixation. *Nano Energy* **2020**, *78*, 105368. [CrossRef]
17. Wang, W.; Zhang, H.; Zhang, S.; Liu, Y.; Wang, G.; Sun, C.; Zhao, H. Potassium-Ion-Assisted Regeneration of Active Cyano Groups in Carbon Nitride Nanoribbons: Visible-Light-Driven Photocatalytic Nitrogen Reduction. *Angew. Chem. Int. Edit.* **2019**, *58*, 16644–16650. [CrossRef]
18. Ju, L.; Bie, M.; Tang, X.; Shang, J.; Kou, L. Janus WSSe Monolayer: An Excellent Photocatalyst for Overall Water Splitting. *ACS Appl. Mater. Interfaces* **2020**, *12*, 29335–29343. [CrossRef]
19. Ju, L.; Bie, M.; Zhang, X.; Chen, X.; Kou, L. Two-dimensional Janus van der Waals heterojunctions: A review of recent research progresses. *Front. Phys.* **2021**, *16*, 13201. [CrossRef]
20. Ju, L.; Qin, J.; Shi, L.; Yang, G.; Zhang, J.; Sun, L. Rolling the WSSe Bilayer into Double-Walled Nanotube for the Enhanced Photocatalytic Water-Splitting Performance. *Nanomaterials* **2021**, *11*, 705. [CrossRef]
21. Zhang, J.; Tang, X.; Chen, M.; Ma, D.; Ju, L. Tunable Photocatalytic Water Splitting Performance of Armchair MoSSe Nanotubes Realized by Polarization Engineering. *Inorg. Chem.* **2022**, *61*, 17353–17361. [CrossRef]
22. Li, X.; Li, Z.; Yang, J. Proposed photosynthesis method for producing hydrogen from dissociated water molecules using incident near-infrared light. *Phys. Rev. Lett.* **2014**, *112*, 018301. [CrossRef] [PubMed]
23. Ju, L.; Bie, M.; Shang, J.; Tang, X.; Kou, L. Janus transition metal dichalcogenides: A superior platform for photocatalytic water splitting. *J. Phys. Mater.* **2020**, *3*, 022004. [CrossRef]
24. Lin, Y.C.; Liu, C.; Yu, Y.; Zarkadoula, E.; Yoon, M.; Puretzky, A.A.; Liang, L.; Kong, X.; Gu, Y.; Strasser, A.; et al. Low Energy Implantation into Transition-Metal Dichalcogenide Monolayers to Form Janus Structures. *ACS Nano* **2020**, *14*, 3896–3906. [CrossRef]
25. Zheng, B.; Ma, C.; Li, D.; Lan, J.; Zhang, Z.; Sun, X.; Zheng, W.; Yang, T.; Zhu, C.; Ouyang, G.; et al. Band Alignment Engineering in Two-Dimensional Lateral Heterostructures. *J. Am. Chem. Soc.* **2018**, *140*, 11193–11197. [CrossRef] [PubMed]
26. Maitra, U.; Gupta, U.; De, M.; Datta, R.; Govindaraj, A.; Rao, C.N. Highly effective visible-light-induced H_2 generation by single-layer 1T-MoS_2 and a nanocomposite of few-layer 2H-MoS_2 with heavily nitrogenated graphene. *Angew. Chem. Int. Ed. Engl.* **2013**, *52*, 13057–13061. [CrossRef]
27. Chaurasiya, R.; Dixit, A.; Pandey, R. Strain-mediated stability and electronic properties of WS_2, Janus WSSe and WSe_2 monolayers. *Superlattices Microstruct.* **2018**, *122*, 268–279. [CrossRef]
28. Jin, C.; Tang, X.; Tan, X.; Smith, S.C.; Dai, Y.; Kou, L. A Janus MoSSe monolayer: A superior and strain-sensitive gas sensing material. *J. Mater. Chem. A* **2019**, *7*, 1099–1106. [CrossRef]
29. Ju, L.; Xu, T.; Zhang, Y.; Shi, C.; Sun, L. Ferromagnetism of $Na_{0.5}Bi_{0.5}TiO_3$ (1 0 0) surface with O_2 adsorption. *Appl. Surf. Sci.* **2017**, *412*, 77–84. [CrossRef]
30. Ju, L.; Dai, Y.; Wei, W.; Li, M.; Huang, B. DFT investigation on two-dimensional GeS/WS_2 van der Waals heterostructure for direct Z-scheme photocatalytic overall water splitting. *Appl. Surf. Sci.* **2018**, *434*, 365–374. [CrossRef]
31. Ju, L.; Liu, C.; Shi, L.; Sun, L. The high-speed channel made of metal for interfacial charge transfer in Z-scheme g–C_3N_4/MoS_2 water-splitting photocatalyst. *Materials Research Express* **2019**, *6*, 115545. [CrossRef]
32. Ma, D.; Ju, W.; Li, T.; Zhang, X.; He, C.; Ma, B.; Lu, Z.; Yang, Z. The adsorption of CO and NO on the MoS_2 monolayer doped with Au, Pt, Pd, or Ni: A first-principles study. *Appl. Surf. Sci.* **2016**, *383*, 98–105. [CrossRef]

33. Wang, Y.; Chen, R.; Luo, X.; Liang, Q.; Wang, Y.; Xie, Q. First-Principles Calculations on Janus MoSSe/Graphene van der Waals Heterostructures: Implications for Electronic Devices. *ACS Appl. Nano Mater.* **2022**, *5*, 8371–8381. [CrossRef]
34. Lee, G.-D.; Wang, C.Z.; Yoon, E.; Hwang, N.-M.; Kim, D.-Y.; Ho, K.M. Diffusion, Coalescence, and Reconstruction of Vacancy Defects in Graphene Layers. *Phys. Rev. Lett.* **2005**, *95*, 205501. [CrossRef] [PubMed]
35. Zhang, S.; Wang, X.; Wang, Y.; Zhang, H.; Huang, B.; Dai, Y.; Wei, W. Electronic Properties of Defective Janus MoSSe Monolayer. *J. Phys. Chem. Lett.* **2022**, *13*, 4807–4814. [CrossRef] [PubMed]
36. Cho, B.; Hahm, M.G.; Choi, M.; Yoon, J.; Kim, A.R.; Lee, Y.-J.; Park, S.-G.; Kwon, J.-D.; Kim, C.S.; Song, M.; et al. Charge-transfer-based Gas Sensing Using Atomic-layer MoS_2. *Sci. Rep.* **2015**, *5*, 8052. [CrossRef]
37. Kou, L.; Frauenheim, T.; Chen, C. Phosphorene as a Superior Gas Sensor: Selective Adsorption and Distinct I–V Response. *J. Phys. Chem. Lett.* **2014**, *5*, 2675–2681. [CrossRef]
38. Qiao, M.; Liu, J.; Wang, Y.; Li, Y.; Chen, Z. $PdSeO_3$ Monolayer: Promising Inorganic 2D Photocatalyst for Direct Overall Water Splitting Without Using Sacrificial Reagents and Cocatalysts. *J. Am. Chem. Soc.* **2018**, *140*, 12256–12262. [CrossRef]
39. Kim, D.; Shin, D.; Heo, J.; Lim, H.; Lim, J.-A.; Jeong, H.M.; Kim, B.-S.; Heo, I.; Oh, I.; Lee, B.; et al. Unveiling Electrode–Electrolyte Design-Based NO Reduction for NH_3 Synthesis. *ACS Energy Lett.* **2020**, *5*, 3647–3656. [CrossRef]
40. Chen, S.; Wang, L.-W. Thermodynamic Oxidation and Reduction Potentials of Photocatalytic Semiconductors in Aqueous Solution. *Chem. Mater.* **2012**, *24*, 3659–3666. [CrossRef]
41. Lv, X.; Wei, W.; Li, F.; Huang, B.; Dai, Y. Metal-Free B@g-CN: Visible/Infrared Light-Driven Single Atom Photocatalyst Enables Spontaneous Dinitrogen Reduction to Ammonia. *Nano Lett.* **2019**, *19*, 6391–6399. [CrossRef]
42. Lv, X.; Wei, W.; Wang, H.; Li, F.; Huang, B.; Dai, Y.; Jacob, T. Nitrogen-free TMS_4-centers in metal–organic frameworks for ammonia synthesis. *J. Mater. Chem. A* **2020**, *8*, 20047–20053. [CrossRef]
43. Zhang, J.; Zhao, Y.; Wang, Z.; Yang, G.; Tian, J.; Ma, D.; Wang, Y. Boron-decorated C_9N_4 monolayers as promising metal-free catalysts for electrocatalytic nitrogen reduction reaction: A first-principles study. *New J. Chem.* **2020**, *44*, 422–427. [CrossRef]
44. Ren, K.; Wang, S.; Luo, Y.; Chou, J.-P.; Yu, J.; Tang, W.; Sun, M. High-efficiency photocatalyst for water splitting: A Janus MoSSe/XN (X = Ga, Al) van der Waals heterostructure. *J. Phys. D Appl. Phys.* **2020**, *53*, 185504. [CrossRef]
45. Deng, S.; Li, L.; Rees, P. Graphene/MoXY Heterostructures Adjusted by Interlayer Distance, External Electric Field, and Strain for Tunable Devices. *ACS Appl. Nano Mater.* **2019**, *2*, 3977–3988. [CrossRef]
46. Wu, Q.; Wang, H.; Shen, S.; Huang, B.; Dai, Y.; Ma, Y. Efficient nitric oxide reduction to ammonia on a metal-free electrocatalyst. *J. Mater. Chem. A* **2021**, *9*, 5434–5441. [CrossRef]
47. Long, J.; Chen, S.; Zhang, Y.; Guo, C.; Fu, X.; Deng, D.; Xiao, J. Direct Electrochemical Ammonia Synthesis from Nitric Oxide. *Angew. Chem. Int. Edit.* **2020**, *59*, 9711–9718. [CrossRef] [PubMed]
48. Bronsted, J.N. Acid and Basic Catalysis. *Chem. Rev.* **1928**, *5*, 231–338. [CrossRef]
49. Evans, M.; Polanyi, M. Inertia and driving force of chemical reactions. *Trans. Faraday Soc.* **1938**, *34*, 11–24. [CrossRef]
50. He, C.-Z.; Zhang, Y.-X.; Wang, J.; Fu, L. Anchor single atom in h-BN assist NO synthesis NH_3: A computational view. *Rare Metals* **2022**, *41*, 3456–3465. [CrossRef]
51. Xiao, Y.; Shen, C. Transition-Metal Borides (MBenes) as New High-Efficiency Catalysts for Nitric Oxide Electroreduction to Ammonia by a High-Throughput Approach. *Small* **2021**, *17*, e2100776. [CrossRef]
52. Niu, H.; Zhang, Z.; Wang, X.; Wan, X.; Kuai, C.; Guo, Y. A Feasible Strategy for Identifying Single-Atom Catalysts Toward Electrochemical NO-to-NH_3 Conversion. *Small* **2021**, *17*, e2102396. [CrossRef]
53. Kohn, W.; Sham, L.J. Self-Consistent Equations Including Exchange and Correlation Effects. *Phys. Rev.* **1965**, *140*, A1133–A1138. [CrossRef]
54. Hohenberg, P.; Kohn, W. Density functional theory (DFT). *Phys. Rev.* **1964**, *136*, B864. [CrossRef]
55. Grimme, S.; Antony, J.; Ehrlich, S.; Krieg, H. A consistent and accurate ab initio parametrization of density functional dispersion correction (DFT-D) for the 94 elements H-Pu. *J. Chem. Phys.* **2010**, *132*, 154104. [CrossRef]
56. Ohnishi, S.T. Measurement of NO Using Electron Paramagnetic Resonance. In *Nitric Oxide Protocols*; Titheradge, M.A., Ed.; Humana Press: Totowa, NJ, USA, 1998; pp. 129–153.
57. Monkhorst, H.J.; Pack, J.D. Special points for Brillouin-zone integrations. *Phys. Rev. B* **1976**, *13*, 5188. [CrossRef]
58. Nørskov, J.K.; Rossmeisl, J.; Logadottir, A.; Lindqvist, L.; Kitchin, J.R.; Bligaard, T.; Jónsson, H. Origin of the Overpotential for Oxygen Reduction at a Fuel-Cell Cathode. *J. Phys. Chem. B* **2004**, *108*, 17886–17892. [CrossRef]
59. Ju, L.; Shang, J.; Tang, X.; Kou, L. Tunable Photocatalytic Water Splitting by the Ferroelectric Switch in a 2D $AgBiP_2Se_6$ Monolayer. *J. Am. Chem. Soc.* **2020**, *142*, 1492–1500. [CrossRef]
60. Mao, X.; Kour, G.; Zhang, L.; He, T.; Wang, S.; Yan, C.; Zhu, Z.; Du, A. Silicon-doped graphene edges: An efficient metal-free catalyst for the reduction of CO_2 into methanol and ethanol. *Catal. Sci. Technol.* **2019**, *9*, 6800–6807. [CrossRef]
61. Li, D.-H.; Li, Q.-M.; Qi, S.-L.; Qin, H.-C.; Liang, X.-Q.; Li, L. Theoretical Study of Hydrogen Production from Ammonia Borane Catalyzed by Metal and Non-Metal Diatom-Doped Cobalt Phosphide. *Molecules* **2022**, *27*, 8206. [CrossRef]

62. Liu, X.; Xu, Y.; Sheng, L. Al-Decorated C_2N Monolayer as a Potential Catalyst for NO Reduction with CO Molecules: A DFT Investigation. *Molecules* **2022**, *27*, 5790. [CrossRef] [PubMed]
63. Ju, L.; Liu, P.; Yang, Y.; Shi, L.; Yang, G.; Sun, L. Tuning the photocatalytic water-splitting performance with the adjustment of diameter in an armchair WSSe nanotube. *J. Energy Chem.* **2021**, *61*, 228–235. [CrossRef]

Disclaimer/Publisher's Note: The statements, opinions and data contained in all publications are solely those of the individual author(s) and contributor(s) and not of MDPI and/or the editor(s). MDPI and/or the editor(s) disclaim responsibility for any injury to people or property resulting from any ideas, methods, instructions or products referred to in the content.

MDPI AG
St. Alban-Anlage 66
4052 Basel
Switzerland
www.mdpi.com

Molecules Editorial Office
E-mail: molecules@mdpi.com
www.mdpi.com/journal/molecules

Disclaimer/Publisher's Note: The title and front matter of this reprint are at the discretion of the . The publisher is not responsible for their content or any associated concerns. The statements, opinions and data contained in all individual articles are solely those of the individual Editor and contributors and not of MDPI. MDPI disclaims responsibility for any injury to people or property resulting from any ideas, methods, instructions or products referred to in the content.